American Statistical Association

1995
Proceedings

of the

Section on Statistical Graphics

Papers presented at the
Annual Meeting of the American Statistical Association,
Orlando, Florida, August 13-17, 1995,
under the sponsorship of
the Section on Statistical Graphics

American Statistical Association ■ 1429 Duke Street, Alexandria, VA 22314

The papers and discussions in this Proceedings volume are reproduced exactly as received from the authors. None of the papers has been submitted to a refereeing process. However, the authors have been encouraged to have their papers reviewed by a colleague prior to final preparation. These presentations are presumed to be essentially as given at the Joint Statistical Meetings in Orlando or other conferences where indicated. The 1995 Proceedings volumes are not copyrighted by the Association; hence, permission for reproduction must be obtained from the author, who holds the rights under the copyright law.

Authors in these Proceedings are encouraged to submit their papers to any journal of their choice. The ASA Board of Directors has ruled that publication in the Proceedings does not preclude publication elsewhere.

American Statistical Association
1429 Duke Street
Alexandria, Virginia 22314

PRINTED IN THE U.S.A.

ISBN 1-883276-28-4

TABLE OF CONTENTS

Invited Papers by Topic

Contributed Papers by Topic

Contributed Papers-Poster Sessions

COMPUTER GRAPHICS IN STATISTICS: THE LAST 30 YEARS IN BRIEF

Sally C. Morton, Dianne Cook, Werner Stuetzle, Andreas Buja

Sally C. Morton, The RAND Corporation, 1700 Main St, Santa Monica, CA 90407-2138
Dianne Cook, Department of Statistics, Iowa State University, Ames, IA
Werner Stuetzle, Department of Statistics, University of Washington, Seattle, WA
Andreas Buja, AT&T Bell Laboratories, Murray Hill, NJ

Keywords. dynamic graphics, grand tour, linked brushing, visualization

1 Motivation

The research area of statistical graphics has a rich and diverse history. The power and originality of many modern graphical techniques cannot be demonstrated fully unless done so dynamically. Thus, researchers frequently have produced videotapes of their methods in action. The Statistical Graphics Section of the American Statistical Association has compiled a lending library of such videos. The library is currently composed of thirty or so videotapes, made as early as 1964, and information on borrowing tapes is provided at the end of this Proceedings entry.

Given the wealth of material in the library, the Statistical Graphics Section proposed that a video introduction and historical overview to this research area be made using clips from the library contents. As a pilot project, a short compilation was constructed and presented at the 1995 ASA Meetings. The video was shown as the first talk in a session on the history of statistical graphics.

2 Video Script

John Tukey opening clip with sound.
We begin this historical video with a clip of John Tukey from the PRIM-9 movie, because the PRIM-9 interactive data display and analysis system, created by Fisherkeller, Friedman, and Tukey in 1973, clearly is the seminal event in the history of statistical graphics. It established statistical graphics as a research

area, and it contained, albeit in rudimentary form, many ideas that were drawn upon in later work.

Our historical video gives an overview of statistical computer graphics research of the past 30 years. There is no attempt to be comprehensive in featuring all the wealth of research developed in these years but rather the material contained in this video highlights what we feel are some of the main conceptual advances in the field. Neither are the selections an attempt to assign priority or credit to individual researchers. Many researchers have made significant contributions to the field of statistical graphics and if you are interested in learning more, the Statistical Graphics Section of the American Statistical Association maintains a lending library of videos demonstrating the spectrum of this research.

The development of statistical graphics was driven to a significant extent by general advances in computer hardware and user interface design. We will point out these computing advances as we show clips from videos demonstrating various statistical graphics systems. PRIM-9 is used as a reference in our history, because PRIM-9 formed a focal point for initiating computer graphics as a research field in statistics.

PRIM-9
Tukey, Friedman, Fisherkeller
1973

PRIM-9 was an interactive computer graphics system for "Picturing, Rotation, Isolation and Masking" up to 9 dimensions. Plots were displayed on an IDI-IOM vector scope driven by a Varian 620 minicomputer. The Varian was connected through a parallel

1

link to an IBM 360/91 mainframe, which performed most of the computations. Controls to the graphical tools were a set of 32 hardware buttons. When PRIM-9 was up and running it so monopolized the computing power of the IBM mainframe that other computing jobs ground to standstill.

The basic ideas for later graphical research were for the most part encompassed in PRIM-9: picturing meant showing two dimensional projections, rotation allowed arbitrary two-dimensional projections rather than variables viewed pairwise, isolation allowed points to be erased so that the analyst could focus attention on subsets of the original data, and masking was effectively conditioning the view based on a third hidden variable. Note that the PRIM-9 display emphasizes the scatterplot as the main plotting method.

It is important to realize that PRIM-9 didn't arise in a vacuum. These next two clips illustrate work done at AT& T Bell Labs.

Chang 1970

Chang's tape displays interactive 3-dimensional rotation of five dimensional data. She used the tool to find one structured 2-dimensional projection embedded in 3 additional dimensions of noise. This attempt predated the rotation shown in PRIM-9 and is a precursor to subsequent techniques for searching high-dimensions for structured projections either automatically or with user control.

Kruskal 1970

Kruskal's tape displays an animation of multidimensional scaling showing the successive repositioning of points to find the layout that best represents their proximity in multivariate space. This system, involving visualization of a statistical algorithm, contrasts an automatic approach with the interactive one employed by Chang and PRIM-9.

McDonald mid-80s

The next major hardware and user interface development can been seen in work by McDonald in the mid-80s. In this clip we see the raster graphics display screen of a single user workstation operating a Window system with three plotting windows visible. The left window displays a geographic image of the region around Manaus, Brazil showing the confluence of two rivers. The upper river is the Rio Negro and the lower one is the Rio Solimões. The top scatterplot shows bands one and two from a satellite and the bottom scatterplot shows bands three and four. This is an example of an implementation of linked brushing between multiple views. Color brushing in one view automatically changes the color of corresponding points in the other views. The realization was done using an object-oriented programming language in an environment where visualization was fully integrated.

Newton 1976

The conceptualization of identifying cases and linked brushing between plots arose in work by Newton in 1976. The catalyst for her ideas was hands-on data analysis. She worked on an IMLAC vector based terminal connected to a mainframe via a serial line, in a set-up closer in taxonomy to that used for PRIM-9 than the single-user workstation used by McDonald.

XGobi
Swayne, Cook, Buja
1991

The last section of our video concentrates on work from the early 90s. XGobi, like many other software packages now available, combines the graphical techniques we have discussed and delivers them into the hands of the masses of UNIX/X Window System users. It provides a one window interface filled with control buttons, scroll bars, and pull down menus for ease of use. Linked brushing and identification are available between multiple windows. The grand tour, conceived by Asimov in 1985, is also implemented in XGobi. This rotation method reflects a return to the original motivation of using motion to view arbitrary projections but it provides a fully representative selection of views, which the Chang and PRIM-9 systems did not. We note that the graphics in XGobi are based on scatterplots, just as in PRIM-9. A wealth of graphical research has been based on univariate plotting methods such as histograms and boxplots which we have not done justice to in this historical compilation.

This grand tour is of the 7-dimensional particle data, which was first analyzed in PRIM-9. The seven colors represent interesting subsets of the data found by an automatic search algorithm which XGobi visualizes for the user. The data takes a simple geometric form.

John Tukey closing clip

3 The Video Lending Library

The Video Lending Library of the ASA Statistical Graphics Section is currently composed of 30 or so videotapes, made as early as 1964. The collection is of great technical and historic interest, and many tapes can be usefully shown to students of statistics. Videos are freely loaned to members of the section, and available for a small fee to nonmembers.

To borrow a video, or to add one to the collection, get in touch with Deborah Swayne by sending email to dfs@bellcore.com or by calling (201) 829-4263.

THE NEW COGNITIVE PERSPECTIVE IN STATISTICAL MAP READING

Douglas Herrmann and Linda Williams Pickle, National Center for Health Statistics
Douglas Herrmann, NCHS, ORM, 6525 Belcrest Road, Hyattsville, Maryland 20782

Key Words: Cognitive theory; statistical maps

Abstract

The National Center for Health Statistics has been conducting an interdisciplinary research program in the human processes of statistical map reading. Our research indicates that the reading of any statistical map is achieved through the performance of four cognitive stages: (1) map orientation, 2) legend comprehension, 3) map /legend integration, and 4) extraction of information from the map. Each cognitive stage concerns a series of psychological processes, e.g., perception, memory, and problem solving that differs between stages. This paper presents a newly developed model of statistical map reading based on these stages. The model is then evaluated in light of recent research on statistical map reading, effects of training and experience on map reading, and problematic map designs in the literature. Finally, the paper addresses the implications of the model for new schemes for pretesting and for map design.

Introduction

The purpose of a good statistical map is to clearly and quickly convey statistical information in a manner that facilitates the map readers' understanding. Maps permit the recognition of patterns in the geographic distributions of statistics that are not readily identified in a table. For data from a large number of small areas, there is no graphical alternative to the map. However, it is no simple matter to design a statistical map that is maximally effective, i.e., easy for statistical map readers to read.

Several years ago we decided to develop to model the cognitive processes involved in statistical map reading in order to develop a coherent approach to statistical map design. At the 1993 ASA meetings we presented a preliminary framework of a model of statistical map reading (Herrmann, Pickle, Kerwin, Croner, White, Jobe, & Jones, 1993). Since that presentation we have conducted considerable research that provides a test of the adequacy of the model's assumptions. Additionally, we have also presented the model in a variety of professional forums and, thereby, have received helpful and constructive feedback about the model from many colleagues at other institutions. This paper begins by presenting the current form of the model and then evaluates the model in light of recent research on statistical map reading, research on the effects of training and experience on map reading, and analyses of common problems in map designs found in the literature. Finally, we discuss how consideration of the cognitive aspects of statistical map reading can lead to improved map design.

Description of the Model

Map-reading factors. The effectiveness of any map design depends upon three factors: map attributes, map reader characteristics, and the map-reading task. The attributes of the map, such as the color scheme and pattern discriminability, impose varying cognitive demands. Different map readers will have different levels of relevant experience and cognitive skills. Finally, map reading tasks differ in their level of complexity and cognitive demands placed on the readers.

Map-reading tasks may be grouped into three broad goals. The most straightforward is to read an approximate rate from the map. If a precise answer is required, using a table of rates would be preferred, but nonetheless it is often of interest to get a rough idea of what the rate is in a specific area. The more common map-reading goal is pattern detection, such as identifying regional patterns, clusters or "hot spots" in the data. When several maps are available, a third goal of pattern comparison is also possible. Most "experts" now agree that a map should be designed with the map-reading task in mind, and that it is difficult to design a map that is ideal for several different types of tasks (Dent, 1993; Monmonier, 1993).

Cognitive subtasks. The model we have developed assumes that any statistical map-reading task can be broken down into *cognitive subtasks*. The value of analyzing statistical map reading in this way lies in the fact that each subtask makes its own peculiar demands on cognition. Instead of conceiving of a map as defective overall, the substask approach assumes that a map may lead to ineffective map reading because of just one or certain subtasks. The subtasks that we believe are essential to statistical map reading overall are delineated in Table 1.

Each subtask is assumed to take time to execute. It is possible that a map reader may perform two subtasks at the same time, although as will be seen below, our research thus far indicates that each subtask is executed separately from the others. The effects of the map orientation and legend comprehension stage may be seen to be crucial to the effectiveness of the integration stage and, in turn, to the effectiveness of the final stage of information extraction.

Each cognitive subtask can be shown to depend on a series of processing stages that differ somewhat from subtask to subtask. The focus of our research concerning the subtask model has been on choropleth maps (Gale &

4

Table 1
Cognitive Subtasks of Statistical
Map Reading

Map Orientation: What does the map represent? What is it designed to communicate?

Legend Comprehension: How is the symbolization scheme meant to be interpreted?

Map/Legend Integration: What are the data values associated with various locations (geographic units)?

Extraction of Information: Which location ranks higher? How does the value for an area compare to a referent value (e.g., U.S. rate)?

Halperin, 1982; Mersey, 1984; Pickle, Mason, Howard, Hoover, & Fraumeni, 1987) for rates. However, we believe that the general theoretical framework may be extended to other kinds of maps for rates or counts.

All four subtasks make use of sensory and preliminary perceptual processes. In the case of *map orientation*, the visual characteristics of different areas must be resolved into patterns of figure and background corresponding to geographic units with different rates. To *understand the legend*, one must be able to identify how many choropleth categories are being used as well as to remember the ordering of the categories. *Integrating* the map with the legend involves detecting visual similarities between the legend and map. Finally, *the extraction of information from the patterns and relationships*, visual properties of a map must be compared.

Support of Recent Research on Statistical Map Reading for the Subtask Model

Considerable research supports the basic subtask framework advanced here (Herrmann & Pickle, in press; Pickle & Herrmann, 1994, Pickle & Herrmann, 1995). First, exploratory research through open-ended interviews with map readers readily demonstrated the subtasks (Beu, Mingay, & White, 1989; Jobe & Beu, 1991; Kerwin & Herrmann, 1992). Readers direct their gaze back and forth between the legend and the map in a manner consistent with the map orientation, legend comprehension, and integration. Moreover, they report having to perform these subtasks and the final subtask of having to extract information.

Second, as suggested above, research has measured the time to complete subtasks and yielded durations consistent with a sensible use of the information contained in each subtask. For example, Hastie, Hammerle, Kerwin, Croner, & Herrmann (in press) found that reading the legend and

becoming oriented to the map took readers (n = 7 college students) approximately 1.64 seconds. After orienting themselves to the map and integrating the legend and map, readers discerned the rate information in about 1.9 seconds. The sum of the inspection time given above (1.64 seconds), and the rate discernment time (1.9 seconds) is very close to the time taken by readers who performed all of these subtasks together (3.50 seconds)..

Third, research also supports the cognitive analysis of the subtasks. For example, the pattern of accuracies and response times changes with task complexity in the study by Hastie et al. (1993). Simple tasks were affected predominantly by perceptual factors. Readers (n = 16 college students) were faster at identifying the rate for a geographic unit if the visual scale was multi-hued than if it was a scale of graduated hue or gray. Detection of clusters (n = 30 college students) also varied across different scales in keeping with their challenge to perception (Lewandowsky, Herrmann, Behrens, Li, Pickle, & Jobe, 1993). Thus, with complex tasks, the time recorded may not accurately reveal the effects of visual map attributes on perception time, because these effects may be masked by the longer time spent in higher level cognitive reasoning, for example, averaging the rates of two or more geographic units (Antes & Chang, 1990).

Support of Recent Research on the Effects of Experience on Statistical Map Reading

Because the model assumes different subtasks, it might be expected that people differ in the skill with which they execute these subtasks. For example, people in certain professions get more experience with graphics than other professions. Thus, an implication of the subtask model is that map readers will differ in map reading skill because of differences in underlying subtask skill and that such differences originate in professional experience. Our research indicates that skill differences do exist across different professions and that these differences occur at a high level of cognition that bears on a map reader's preference for reading maps of different styles. For example, Maher (1992) found in a study of 10 epidemiologists, 10 statisticians, and 10 congressional staffers that statisticians verbalized what they read from a map less than epidemiologists did, and congressional staffers verbalized even less than statisticians. The results also suggested that statisticians preferred complex maps more than epidemiologists and epidemiologists preferred complex maps more than the congressional staffers. Similarly, White, Croner, Herrmann, and Pickle (1994) found that statisticians preferred to read maps of a different style than did a variety of other professionals (e.g., epidemiologists, psychologists, computer programmers).

Problematic Map Designs in the Literature

Numerous articles and some books have catalogued a variety examples of maps whose design thwarted understanding of the data in the map. Our examination of many of these examples of problematic map designs almost invariably reveals that the defect that thwarts understanding affects a particular subtask or subtasks (Croner, Herrmann, Pickle, & White, 1995). For example, many maps use confusing legends. A sensible legend arrangement places low values at the bottom of the legend and high values at the top. Also, a sensible legend uses white or light shadings for zero and darkness for increased magnitude. Maps that use legends with low values at the top and/or light shadings at the top often result in misreading because this ordering goes against most map reader's expectations for legend design.

To illustrate the dependence of map design on the primary map-reading task, the maps below (Figure 1) depict the percent change in the teenage pregnancy rate by state (CDC 1993). The left map, adapted from the originally published map, uses six very distinct patterns for the six categories (plus white for missing data), whereas the right map uses a gradation of shades, representing greater positive change by darker shades. One may choose either based on aesthetic factors, but neither map is the optimum design for both rate-readout and pattern identification. For rate-readout tasks, the distinctive hatch patterns on the left map facilitate matching the state pattern to the legend; the category is not so readily discernible from the right map because of the similarities of the gray shades. On the other hand, the visual similarity of states with similar percentages on the right map facilitates the identification of regional patterns or state clusters. Thus, a map design may be poor for a particular task, but optimum for another task.

Map-reading factors (i.e., map attributes, map-reader characteristics, task) can affect one or more of the four cognitive subtasks of the proposed model. For example, map orientation is easiest when the map reader can easily perceive the boundaries of the geographic units of concern. The absence of boundaries, boundaries that are difficult to see, or boundaries of a geographic area unfamiliar to the reader can interfere with map orientation. Similarly, integration of the map with the legend is made difficult by the same factors that interfere with legend comprehension and map orientation. Obviously, factors that interfere with the preceding three stages make highly unlikely that the appropriate information is extracted from the map in the final stage.

Discussion

The subtask model has turned out to be very helpful to us in both research and map design (Sirken, Herrmann, & White, 1993). Presently, the variety of visual representation schemes is so great that different maps of the same data are often seen as reflecting different data bases (Pickle, Herrmann, Kerwin, Croner, & White, 1993). The results of our research will help elucidate why such variation can occur and provide knowledge to professionals who produce maps so as to ensure that maps are accurately understood.

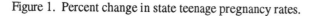

Figure 1. Percent change in state teenage pregnancy rates.

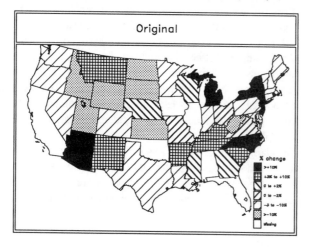
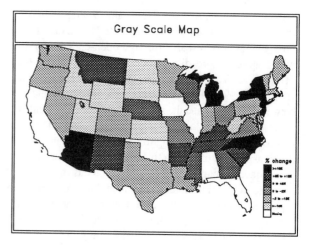

References

Antes, J. R., & Chang, K. (1990). An empirical analysis of the design principles for quantitative and qualitative area symbols. *Cartography and Geographic Information Systems*, 17, 271-277.

Beu, D. H., Mingay, D. J., & White, A. A. (1989). Cognitive experiments in data presentation. *Proceedings of the Statistical Graphics Section, American Statistical Association*, 30-35.

Centers for Disease Control (1993). Teenage pregnancy and birth rates - U.S. 1990. *Morbidity Mortality Weekly Report*, 42, No. 38, 735.

Croner, C. M., Herrmann, D. J., Pickle, L. W., and White, A. A. (1995). A cognitive model of statistical map reading: Implications for GIS. Submitted for publication.

Dent, B. D. (1993) *Cartography: Thematic Map Design*. Dubuque, Iowa: Wm. C. Brown Publishers.

Gale, N., & Halperin, W. C. (1982). The case for better graphics: The unclassed choropleth map. *The American Statistician*, 36, 330-336.

Hastie, R., Hammerle, O., Kerwin, J., Croner, C., & Herrmann, D. J. (In press). Data-display format compatibility and human performance reading statistical maps. *Journal of Experimental Psychology: Applied*.

Herrmann, D. & Pickle, L. W. (In press). A cognitive subtask model of statistical map reading. *Visual Cognition*.

Herrmann D, Kerwin J, Pickle LW, Croner C, White A, Jones G, Jobe J: Cognitive processes in statistical map reading. *Proceedings of the Statistical Graphics Section, American Statistical Association 1993 Meeting*, San Francisco, CA. Pp. 51-54.

Jobe, J. B., & Beu, D. H. (1991). Cognitive factors in interpreting statistical graphs. Unpublished manuscript. Hyattsville, MD: NCHS.

Kerwin, J., & Herrmann, D. (1992). Benefits of pretesting statistical maps with cognitive interviews. Unpublished manuscript. Hyattsville, MD: NCHS.

Lewandowsky, S., Herrmann, D., Behrens, J. T., Li, S., Pickle, L., & Jobe, J. B. (1993). Perception of clusters in statistical maps. *Applied Cognitive Psychology* 7:533-551, 1993.

Maher, R. (1992). Statistical map perception as a function of expertise for congressional staffers, epidemiologists, and statisticians. Unpublished manuscript. Hyattsville, MD: NCHS.

Mersey, J. E. (1984). Colour and thematic map design: The role of colour scheme and map complexity in choropleth map communication. *Cartographica*, 27, Monograph 41.

Monmonier, M. (1993). *Mapping It Out*. Chicago: The University of Chicago Press.

Pickle, L. W., & Herrmann, D. (1994). The process of reading statistical maps: The effect of color. *Statistical Computing and Graphics*, 5, 1, 9-16.

Pickle, L. W. & Herrmann, D. (1995). Cognitive Aspects of Statistical Mapping. *Cognitive Methods Staff Working Paper Series, No. 18*, National Center for Health Statistics, Hyattsville, Maryland.

Pickle, L. W., Herrmann, D., Kerwin, J., Croner,C. M., & White, A. A. (1993). The impact of statistical graphic design on interpretation of disease rate maps. *Proceedings of the Statistical Graphics Section, American Statistical Association*. Pp. 51-54.

Pickle, L. W., Mason, T. J., Howard, N., Hoover, R., & Fraumeni, J. F., Jr. (1987). *Atlas of U.S. cancer mortality among whites: 1950-1980*. DHHS (NIH) Publ.. No. 87-2900. Washington, DC: U.S. Government Printing Office.

Sirken, M., Herrmann, D., & White, A. (1993). The cognitive aspects of map research at the National Center for Health Statistics. *Proceedings of the Government Statistics Section, American Statistical Association*.

White AA, Pickle LW, Herrmann DJ, Croner CM, Wilson BF (1994). Map design preferences associated with professional discipline. *Proceedings of the Statistical Graphics Section, American Statistical Association*, Toronto, Canada. Pp. 54-59.

VISUAL DETECTION OF CLUSTERS IN STATISTICAL MAPS

Stephan Lewandowsky, University of Western Australia, John T. Behrens, Arizona State University
Stephan Lewandowsky, Department of Psychology, University of Western Australia,
Nedlands, W.A. 6907, Australia (lewan@psy.uwa.edu.au)

Key Words: Statistical maps, Visual perception

Abstract

The visual detection of clusters in statistical maps is an important analysis tool in epidemiology and public health research. This article surveys some of the relevant psychological literature to provide guidelines for the construction of statistical maps that are particularly suitable for cluster detection.

Introduction

The visual display and analysis of data has become increasingly important in many areas of research. In parallel, there has been increasing recognition among statisticians that graphing of data is a non-trivial exercise whose success relies on knowledge of people's perceptual capabilities and limitations. Several recent experimental investigations by statisticians (e.g., Cleveland & McGill, 1984) and psychologists (e.g., Lohse, 1993) have provided the first elements of an empirical data base that can guide the construction of "user-friendly" graphs. (For a review, see Lewandowsky & Spence, 1990).

Perhaps most challenging of all graphs is the statistical map which differs from other data displays in several important ways: First, the two dimensions of the plane are necessarily devoted to representing geographical information, thus eliminating use of the perceptually most directly interpretable codes (position along the horizontal and/or vertical axis) for data representation. By implication, simultaneous mapping of more than one statistical variable presents an even more pernicious problem. Second, it is not unusual for a map to show many hundreds or indeed thousands of data points, for example when a variable is represented at the level of U.S. counties. Third, when a variable is represented by shading or coloring of geographic regions, the perceived magnitude of each data value is confounded with the, typically unrelated, size of the corresponding geographic region. Finally, the use of hue to represent magnitude requires awareness of the perceptual complexity of color. Subjective perception of color is governed by intricate interactions between brightness, saturation, and hue, which, when left uncontrolled, can engender serious misperceptions and false interpretations.

Notwithstanding these constraints and limitations, statistical maps, especially those representing mortality data, have had a unique influence on public policy throughout history. The earliest known case involved mapping of cholera deaths in London in 1854, which allowed the water-borne illness to be traced to a contaminated water pump (discussed in Gilbert, 1958; Maher, 1995a; Wainer, 1992). More recently, publication of the U.S. Cancer Atlas (Pickle, Mason, Howard, Hoover, & Fraumeni, 1987) identified a particularly high incidence of cervical cancer in West Virginia, which prompted state legislators to allocate extra funds towards early detection and treatment of this often curable disease, with the result that mortality subsequently declined (Maher, 1995a).

In these cases, as in the majority of epidemiological applications of mortality maps, emphasis was on the detection of "clusters", contiguous regions of particularly high (or low) mortalities that are of interest because they represent abnormal or unusual situations. Once clusters are detected, corrective efforts can be applied immediately (as in the case of West Virginia) or analysis can focus on discovering epidemiologically relevant correlates (such as a contaminated water pump). Visual analysis is often the preferred way to detect clusters because purely mathematical approaches are limited and cannot model all relevant features of geographical space (Marshall, 1991).

The purpose of this article is to provide guidelines for the construction of statistical maps that are primarily intended to aid in the detection of clusters. In the absence of such guidelines, examination of some 50 mortality atlases revealed minimal consistency (Walter & Birnie, 1991). In particular, although hue was used to code data in more than half of the atlases examined by Walter and Birnie (1991), four colors (orange, brown, blue, and green) were variously used to represent each possible numeric category—from highest to lowest. To help resolve this unnecessary variability, we first briefly describe existing mapping techniques and discuss some a priori considerations of map construction. We then survey the relevant psychological research on cluster detection to compile empirically-supported guidelines for the design of maps.

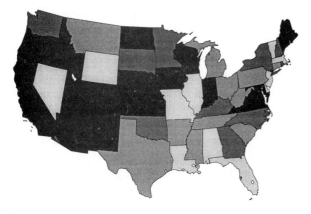

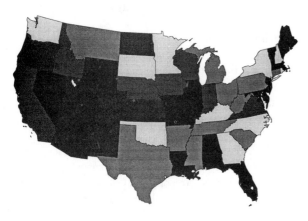

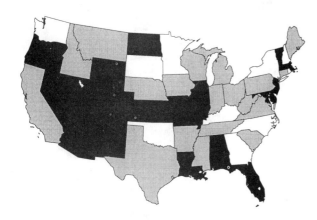

Statistical Maps

Statistical maps can be broadly classified into three categories corresponding to different methods of data coding: Choropleth, density, and symbol maps.

Choropleth maps use shading or coloring of individual areas on the map to represent magnitudes. Most choropleth maps are classed; that is, data values are aggregated into few (typically 5-8) class intervals that sub-divide the overall range of magnitudes. The three maps shown in the figure on the left are examples of classed choropleth maps. *Monochrome* choropleth maps use increasingly dense shadings or fills of a single color to represent increasing numerical magnitude. *Double-ended* choropleth maps, also known as *bipolar* or *two-opposing colors* maps, use increasingly dense shadings of one color to represent high (above-midrange) magnitudes and increasingly dense shadings of another color to represent low (below-midrange) magnitudes. *Categorical color* or *multi-hue* maps, finally, assign a different fully-saturated hue to each class of magnitudes. Common to all choropleth maps is the fact that a given numerical magnitude will be given greater visual prominence if it happens to fall into a larger region because a greater map area will be devoted to shading or coloring.

Density maps represent magnitudes by placing numerous symbols, usually simple dots, into each map area. Increasing magnitudes thus lead to an increased visual density of plotting symbols. In contrast to choropleth maps, a given numerical magnitude will be visually enhanced if it falls into a *smaller* area because the same number of dots will be crowded into less space.

Symbol maps, finally, represent magnitudes by superimposing some graphical symbol, such as a pie chart or a framed rectangle, onto each map area. In consequence, perception of symbol maps is not susceptible to areal bias because the symbol size is usually unaffected by the size of map areas—unless, of course, those areas are smaller than the printed symbol.

A Priori Considerations

The relative merits of these various maps are best adjudicated by psychological experimentation. Nonetheless, some design decisions also need to be informed by a priori considerations.

For example, the publication of color-coded maps carries the considerable risk that information will be lost in black-and-white photocopies of the original. Similarly, it must be remembered that roughly 7% of males are colorblind and thus unable to differentiate between certain sets of hues. Lest one think this is a minor point, the reader is invited to photocopy the cancer maps in Boyle, Muir, and Grundmann (1989), most of which lose all intelligibility in black and white. As a further example, consider the maps in the figure on the left which represent mortality rates for the same disease (age adjusted death rates for white males from asthma during 1986-88) using monochrome reproductions of color originals. For the top map, hues were assigned to magnitudes in order of increasing brightness, thus ensuring that the data are conveyed accurately even in monochrome reproductions. Comparison with the remaining two maps, which use a spectral categorical arrangement of hues and a red-blue double-ended scale, respectively, reveals the potential difficulties of relying exclusively on color to communicate magnitude.

On these grounds alone, caution is advised when using color coding. Carswell, Kinslow, Pickle, and Herrmann (1995) explored one solution to the reproduction problem by providing double-ended scales that continued to be interpretable on the basis of brightness alone, analogous to the top map on the preceding page.

A related issue concerns the need for legends that arises from the use of color to code magnitude. Specifically, whereas *ordinal* understanding of a monochrome choropleth map does not require presence of a legend, a categorical or double-ended scale is meaningless without a legend. The requirement for a legend may not be an obstacle when the map is intended for detailed inspection of a single data set, but legends may be a nuisance when numerous maps are inspected in order to conduct a quick exploratory analysis of a number of data sets.

Representing Mortalities: Empirical Findings

We first survey the relevant literature concerning the simplest possible case, when all available data for a single variable are to be mapped.

Symbol Maps

Detection of a cluster is contingent upon the observer perceiving contiguous map regions as a visually coherent group. For symbol maps, this requires the perceptual integration of several distinct stimuli, a process that is subject to a trade-off between symbol size and overlap. Specifically, when data must be shown for small geographic regions, adjacent symbols will necessarily overlap unless their size is drastically reduced—in either case, legibility and perceptual integration may be impaired, as illustrated in the figure on the right.

Empirical support for the difficulties associated with symbol maps was provided by Lewandowsky, Herrmann, Behrens, Li, Pickle, and Jobe (1993). They found that pie maps, similar to those shown in the figure on the right, were judged consistently to contain *fewer* clusters than a variety of choropleth maps displaying the same mortality data.

Density Maps

Use of the density map for visual cluster detection is associated with a blend of advantages and problems. On the more negative side, it is known that numerosity estimates for arrays of randomly placed dots are affected by the perception of subjective clusters (e.g., van Oeffelen & Vos, 1982). The obvious implication, that

density maps may give rise to the perception of fortuitous subjective clusters, was confirmed in the study by Lewandowsky et al. (1993), who found that observers tended to perceive clusters in widely differing locations. This perceptual variability was particularly great when observers had to identify clusters of *low* mortalities and the data were coded at the state level, a task that required differentiation between arrays of relatively few and loosely spaced dots. Conversely, on the more positive side, when the data were represented in smaller geographic units, equivalent roughly to counties, Lewandowsky et al. (1993) showed the density map to be competitive with choropleth schemes. Overall however, the sensitivity of the density map to size of geographic unit may render its use inadvisable except perhaps for the most tentative exploratory data analysis.

Choropleth Maps

There is empirical support for every one of the remaining three choropleth maps. Nonetheless, there are significant psychological differences between them that merit discussion. In particular, there has been much debate about the suitability of using hue to represent magnitude.

Double-Ended vs. Monochrome Scales. Turning first to the double-ended scale, the case for color has been stated very eloquently by Carswell et al. (1995). Given that humans typically cannot differentiate more

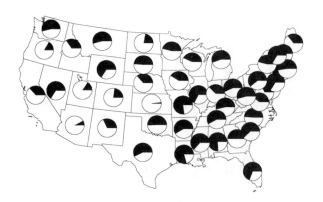

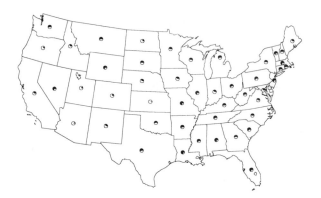

than 3-5 levels of gray or brightness (Sanders & McCormick, 1993), the double-ended scale provides an opportunity to virtually double the number of available categories through concatenation of two hues, each with its own set of shadings. In addition, because low and high magnitudes are typically represented by identical levels of saturation, the double-ended scale affords the opportunity to correct the known human bias of attending primarily to the confirming presence of critical information (i.e., high mortalities), even though its absence (i.e., low mortalities) may be equally relevant to understanding the etiology of a disease.

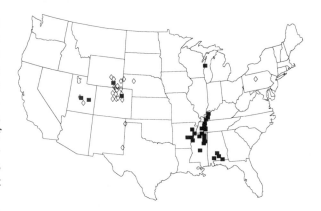

On the other hand, valid psychological reasons can also be cited *against* the double-ended scale, in particular if one assumes that preattentive processes engaged in visual search also underlie the cognitively more elaborate domain of map inspection. There is widespread agreement among cognitive psychologists that visual detection of a target is qualitatively more difficult and time-consuming if conjunctions of features must be formed (e.g., Treisman, 1991). For example, if the features 'red' and 'triangle' must be detected and combined to identify a target—because red squares and green triangles are also present—a qualitatively different perceptual process appears to be engaged than when the task is to pick out a red object among green distractors or a triangle among squares. If these effects also apply to the visual inspection of maps, it follows that a monochrome scale may be more readily perceived than the double-ended scale, which relies on formation of the conjunction of saturation and hue.

A further potentially relevant property of feature conjunctions is that, for stimuli presented outside the immediate spatial focus of attention, *illusory* conjunctions may occur, in which attributes are erroneously exchanged across stimuli (Treisman & Schmidt, 1982). For example, when presented with a pink T and a yellow S, participants may report seeing a pink S instead. In the context of the double-ended scale, illusory conjunctions might manifest themselves as scale-reversals; confusions of high and low mortalities.

The available empirical evidence mirrors the theoretical ambivalence surrounding the double-ended scales. Consider first the more negative results. In the study by Lewandowsky et al. (1993), participants had to simultaneously identify and mark high and low mortality clusters. The double-ended scale was found to lead to somewhat greater perceptual variability (i.e., less agreement among participants about cluster locations which suggests that the map communicates the data less consistently) than two monochrome scales involving shades of blue and black. At least in part, that result was due to the non-negligible number of scale reversals for the double-ended scale, exactly as one

might expect from the visual search literature. The observed scale reversals are illustrated in the figure above: Solid squares indicate the perceived location of high mortality clusters, whereas open diamonds show the perceived locations of low mortality clusters. Even though the map omits the underlying mortality data, it is clear that four (out of 31) participants perceived a high mortality cluster in regions that were most often identified has having the lowest incidence of the disease. These observations are best interpreted as scale-reversals.

Perceptual difficulties associated with the double-ended scale are not confined to a single study: Mersey (1990) also reported occasional scale reversals among participants, and Cuff (1973) found that participants often chose the lightest class interval—that is, the *midpoint* of the double-ended scale—when asked to identify the region with the *lowest* data value. Although each of these studies in isolation is subject to limitations arising from the choice of participants or details of the design (Carswell et al., 1995), there can be little doubt overall that double-ended scale reversals may occur in a non-negligible number of cases.

Turning to the more supportive evidence, Carswell et al. (1995) conducted a particularly thorough investigation of several different double-ended schemes in comparison to a gray scale. Using a variety of measures, including a perceptual variability index akin to Lewandowsky et al. (1993), the results showed that performance with a red-yellow opposing scale was consistently better than with a monochrome gray scale. On the other hand, a blue-yellow opposing scale was found to be inferior to shades of gray, with a third double-ended scale, red-blue, taking up the middleground.

Fortunately, Carswell et al. (1995) were able to provide principled explanations for the observed differences between color pairings, which otherwise would have formed a confusing array of seemingly arbitrary outcomes. Carswell et al. (1995) suggested that the success of the red-yellow scale was in large part due to the decoupling of perceived brightness and saturation. Specifically, for that scale, brightness increased monotonically from the lowest (saturated red) to the highest category (saturated yellow), thus providing an additional perceptual cue for estimation of magnitudes and also eliminating the problems otherwise associated with black-and-white reproduction. The failure of the blue-yellow scale at first appears puzzling because it, too, decoupled brightness and saturation. Carswell et al. (1995) suggested that the blue-yellow scale failed because of its conflicting use of two opposing graphical conventions; 'darker-for-more' (shades of blue) and 'warmer-for-more' (shades of yellow).

Finally, the results of Carswell et al. (1995) and Lewandowsky et al. (1993) deserve to be contrasted for the red-blue scale shared by both studies. The comparison reveals negligible differences in outcome for maps drawn at the state level, although Lewandowsky et al.'s (1993) results tended to favor monochrome maps somewhat more than Carswell et al.' (1995). The pattern appears to change when smaller geographic units (Health Service Areas; comparable in size to counties) are considered, which were used by Lewandowsky et al. (1993) but which unfortunately were absent from the Carswell et al. (1995) study. For those smaller geographic units, Lewandowsky et al. (1993) found performance to be consistently worse with the double-ended scales than monochrome maps. Unfortunately, it must remain unclear whether the red-yellow scale favored by Carswell et al. (1995) would have eliminated the disadvantage for the double-ended map.

On balance, the available experimental results are best summarized as follows:

(1) Double-ended scales that decouple brightness and saturation are highly suitable for cluster detection. Those scales also circumvent the problems associated with black-and-white reproduction and color blindness (Carswell et al., 1995).

(2) Other double-ended scales are at best comparable to monochrome scales and in those cases little appears to be gained by the use of color (Carswell et al., 1995; Lewandowsky et al., 1993).

(3) The occurrence of scale reversals with a double-ended scale cannot be ruled out (Cuff, 1973; Lewandowsky et al., 1993; Mersey, 1990). Particular care must be taken to avoid use of scales that pit graphical conventions against each other (Carswell et al., 1995).

(4) Double-ended scales most often require the presence of a legend. The legend must show values in a vertical orientation, with magnitudes decreasing from top to bottom (Pickle, Herrmann, & Wilson, 1995).

(5) Mapping of data in smaller geographic units may favor monochrome maps over double-ended scales (Lewandowsky et al., 1993).

Categorical Color Scales. Unlike monochrome scales, sets of different colors are not immediately perceived as an ordinal sequence of magnitudes. When hue is uncorrelated with brightness, neither convention nor perceptual principles imply that, say, green map areas somehow represent 'more' (or 'less') than blue or red regions. In consequence, for most categorical color scales, hues are assigned to magnitudes either arbitrarily or in spectral sequence. The evidence shows that, for a given set of colors, assignment to magnitudes has no effect on performance in a variety of tasks (Hastie, Hammerle, Kerwin, Croner, & Herrmann, in press).

However, assignment to magnitudes aside, the choice of colors is an important determinant of performance. For example, Cleveland and McGill (1983) showed that the number and size of red areas was consistently overestimated in comparison to areas filled in green. Similarly, using a cluster detection task, Lewandowsky et al. (1993) found that performance with a categorical color scale differed widely, depending on which set of colors observers had to attend to. Participants had to identify clusters of low and high mortalities on the same map, and for low mortalities, performance with the categorical scale was identical to monochrome maps, whereas for high clusters it was significantly worse. Given the particular hue-to-magnitude assignment used by Lewandowsky et al. (1993), the outcome associates good performance with areas that were predominantly blue and green (low magnitudes) and less satisfactory performance with orange and red areas (high clusters). Because performance did not differ between low and high clusters for any of the other maps, it stands to reason that the effect observed with the categorical scale reflected inherent differences in discriminability between the colors that represented low and high magnitudes.

This hypothesis can be supported by inspection of the color circle in the figure below, which represents a

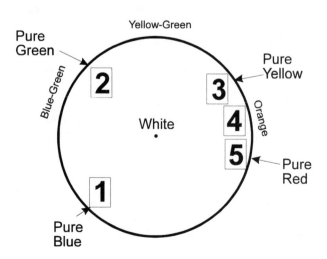

12

convenient way to summarize human color perception (e.g,. McBurney & Collings, 1977, p. 44). In general, discriminability between colors corresponds to the distance that separates them along the circle. The bold numbers in the figure represent the ordinal arrangement, by magnitude, of the colors used by Lewandowsky et al. (1993). It is readily apparent why judgments involving categories 1 and 2 would be more accurate than those involving 4 and 5.

When constructing categorical color scales, it is thus advisable to refer to a color circle, or the more complete color *solid* which also takes account of brightness, to maximize discriminability among categories of magnitude. The data suggest that when this is done, categorical color scales are a very suitable means for mapping of mortality data however, the utility of the scales is seriously limited when discriminability between hues is reduced.

Monochrome Scales. The preceding two sections often referred to monochrome scales for comparison purposes. For ease of reference, these comparisons are summarized here:

(1) Monochrome maps support more consistent cluster detection than certain sub-optimal double-ended scales (Carswell et al., 1995; Lewandowsky et al., 1993). However, performance with monochrome maps does not match performance with judiciously chosen double-ended scales that decouple brightness and saturation (Carswell et al., 1995).

(2) Monochrome maps outperform categorical scales with limited discriminability (Lewandowsky et al., 1993). Monochrome scales remain on par even with judiciously chosen categorical color scales for cluster detection (Lewandowsky et al., 1993) and related tasks (Hastie et al., in press).

(3) The choice of hue for monochrome shading seems to matter little, although to date empirical comparison has involved only red, blue, and black or gray (Hastie et al., in press; Lewandowsky et al., 1993).

Overall, the data on cluster detection suggest that little, if anything, is lost if color is forgone in favor of monochrome maps. However, monochrome scales have at least one significant limitation that becomes apparent when pure magnitude information is to be augmented by indications of statistical reliability.

Representing Mortality and Statistical Reliability

In many situations, the statistical robustness of the data is of prime importance. Public health measures tend to be very costly, and decisions about resource allocations cannot be taken lightly. In consequence, it is often desirable to augment pure magnitude information with indications of statistical reliability or significance. Reliability here refers to whether or not the observation for an area differs significantly from some suitable expected value (see Pickle et al., 1987, for a brief discussion of computational issues). For example, in mortality data, a numerically high risk of death may not reach statistical significance because of a small absolute number of cases in a sparsely populated county. Conversely, a larger population that gives rise to a greater number of deaths may differ significantly from the expected risk even though the mortality rate is only slightly elevated.

To date, practitioners have followed one of two main approaches to the reliability problem. At the one extreme, only those data were mapped that either differed significantly from the map average or fell into the extreme deciles, whereas all remaining areas were left blank (Pickle et al., 1987). This approach prevents attention—and epidemiological effort—from being focused on regions whose mortalities may have been spuriously high. However, clusters arising from statistically unreliable data may nonetheless turn out to be important and worthy of attention, for example if they re-occur across several reporting periods. Hence, an alternative approach to mortality mapping has not distinguished between reliable and unreliable reporting regions (e.g., Wagner, 1989), thus giving equal credence to all observations. The maps discussed in the preceding section followed this approach, which is often found in mortality atlases: Walter and Birnie (1991) reported that, of 62 atlases examined, 32 mapped the data without regard to significance.

A third, inherently preferable approach would seek to show statistical significance together with magnitude information, for example by representing two layers of information on the same map in a bivariate representation (as used by roughly a quarter of the atlases examined by Walter and Birnie, 1991). A number of studies have recently explored this solution, and have compared its effectiveness to the first two approaches of either leaving unreliable areas blank or ignoring reliability altogether.

The Utility of Reliability Representations

Simultaneous mapping of two variables, such as reliability and magnitude, constitutes a non-trivial design challenge because it constrains the available choice of map scales. For example, monochrome scales are largely unsuitable for bivariate representations. Monochrome maps afford representation of a second variable primarily through superimposition of symbols, such as asterisks to denote significance. In consequence, the problems outlined earlier in the context of symbol maps apply also to monochrome maps in bivariate applications. Similarly, bivariate schemes that orthogonally combine two double-ended scales (e.g., a red-yellow and a blue-yellow scale factorially crossed) have been shown to be of dubious perceptual value (Wainer & Francolini, 1980). Given these problems and precedents, it is thus not immediately certain that observers can make use of simultaneous representation of reliability and magnitude.

This issue was examined by Lewandowsky, Behrens, Pickle, Herrmann, and White (1995) in a comparison of cluster detection performance across four variants of a categorical color map. For two of the maps, the visual salience of regions with unreliable data was reduced, either by lowering saturation of the corresponding color or by overlaying the region with a white cross-hatch pattern. The remaining reliable areas were shown in full saturation of the appropriate hue, following standard practice. A third, reliable-only map also followed standard practice for reliable regions but omitted unreliable data by leaving the corresponding areas blank. A control map, finally, showed all areas in fully saturated hues, thus following the approach of presenting all data without regard to reliability.

The results were unequivocal: The variance between perceived cluster locations was twice as great with the map containing blank areas than with the control map displaying all data. The remaining two maps that combined magnitude and reliability fell in-between those extremes, suggesting that there is a slight cost associated with processing of the additional information.

The clear implication that unreliable areas should not be left blank, was confirmed in a further study by Lewandowsky and Behrens (1995) which examined absolute accuracy of cluster detection using artificial mortality data constructed from a known statistical model. Again, when unreliable areas were left blank, performance was impaired in comparison to maps that showed unreliable data and identified them as such.

Bivariate Coding of Reliability.

Having demonstrated the general perceptual utility of bivariate representations, we now compare four competing techniques that have been proposed to map reliability in addition to magnitude. Concerning categorical color maps, unreliable areas have been variously overlayed with a cross-hatch pattern (Lewandowsky et al., 1995; Lewandowsky & Behrens, 1995; MacEachren & Brewer, 1995), reduced in saturation (Lewandowsky et al., 1995), or changed to a lighter hue (MacEachren & Brewer, 1995). The evidence tends to favor cross-hatching over changes in saturation or color, both in terms of performance and subjective preferences.

A fourth coding scheme involved a red-blue double-ended scale in which significance, irrespective of

magnitude, was represented by the two most extreme categories (fully saturated red or blue), and in which unreliable data were represented by the intervening categories. Lewandowsky and Behrens (1995) found that this scheme compared favorably to cross-hatched categorical maps, suggesting that it may be a highly suitable choice for bivariate representation. At least in part, this may reflect the similarity between that double-ended scale and the control maps used by Lewandowsky et al. (1995), in which all areas were represented as reliable and which also supported the most consistent cluster detection performance.

Separate Representations

A different approach to reliability mapping involves simultaneous presentation of two maps, one with magnitude information and a separate one with reliability information. MacEachren and Brewer (1995) compared some of the earlier bivariate maps to pairs of separate maps on a variety of experimental tasks. The bivariate maps resembled the categorical color scales used by Lewandowsky et al. (1995), with unreliable areas identified by a change in hue or cross-hatching. The separate map pairs consisted of a choropleth map for magnitude information that was shown together with a smaller binary (light/dark) monochrome map for reliabilities.

When participants had to compare reliability of the data between census regions, MacEachren and Brewer (1995) found performance to be considerably better with the separate map pairs, presumably because participants could focus on a simple univariate map for these judgments. Similarly, map pairs resulted in significantly more accurate identification of the region with the lowest proportion of reliable data. However, when judgments involved all data on the map, for example when overall reliability had to be assessed, map pairs engendered higher reliability judgments than single bivariate maps, presumably because subjects focused on the magnitude map and found the separate reliability map easier to ignore.

Overall, the results of MacEachren and Brewer (1995) suggest that judgments that focus only on reliabilities are best supported by separate representations, whereas joint considerations of magnitude and reliability are best supported by a judiciously designed bivariate map.

On balance, the available data on representation of reliability can be summarized as follows:

(1) There is clear evidence, based both on consistency and accuracy of perception, that unreliable data should not be omitted from statistical maps (Lewandowsky & Behrens, 1995; Lewandowsky et al., 1995).

(2) Simultaneous bivariate mapping is suitable using either categorical-color or double-ended scales (Lewandowsky & Behrens, 1995; Lewandowsky et al., 1995; MacEachren & Brewer, 1995)

(3) For categorical color maps, unreliable areas are best shown by overlay of a cross-hatch pattern rather than through a change in color or reduced saturation (Lewandowsky et al., 1995; MacEachren & Brewer, 1995).

(4) A double-ended scale that uses two additional extreme categories to represent reliability supports particularly accurate cluster detection (Lewandowsky & Behrens, 1995).

(5) Map pairs constitute a viable alternative that has advantages if the observer focuses primarily on reliability (MacEachren & Brewer, 1995).

The Role of Observer Expertise

Most of the studies reviewed so far involved undergraduate participants with little if any previous experience with statistical maps. One might therefore be legitimately concerned about the generalizability of the results to the more expert audiences that are likely to be exposed to published mortality maps.

Fortunately, although only a few systematic comparisons between experts and novices exist, there have been no reports of large and pervasive performance differences. Maher (1995b) compared the performance of statisticians, epidemiologists, and Congressional staffers on seven measures of map reading performance, including cluster detection. The only reliable difference to emerge was that epidemiologists and statisticians mentioned more relevant detail when describing titles and legends of maps than the Congressional staffers.

Similarly, the study on reliability mapping by Lewandowsky et al. (1995) included two groups of experts, epidemiologists and geographers, whose performance differed from that of novices only when maps were differentiated by very subtle features in the data. The overall pattern of results was identical across all subjects regardless of expertise.

Finally, the examination of double-ended scales by Carswell et al. (1995) also included a comparison between experts and novices. In most cases the groups did not differ, with the exception that the experts were particularly adversely affected by the double-ended scale (blue-yellow) that pitted two graphical conventions against each other.

Overall, the limited available data confirm that for most map perception tasks, the performance of novices differs little, if at all, from that of statistical experts. Only when very subtle features in the data become important, or when the map violates extensive previous

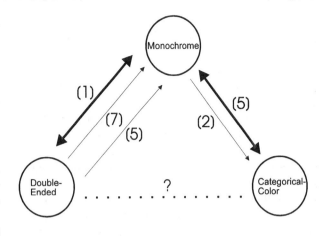

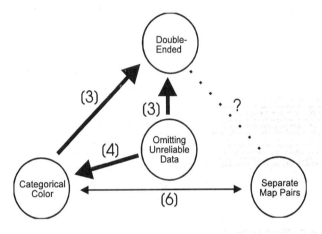

experience, do experts differ from novices. By extension, there is every reason to assume that major design principles uncovered and tested using novice undergraduates will also be relevant to expert audiences.

Conclusions

The two figures above summarize the evidence reviewed here. The top figure refers to maps in which all available data are displayed, and the bottom figure refers to magnitude-and-reliability maps. Each empirical comparison between a pair of maps is represented by an arrow pointing at the favored map, with the thickness indicating the judged strength of the evidence (i.e., magnitude of the performance difference). The alphanumeric labels for each arrow identify the corresponding study in the References. The figures are intended to provide a concise empirical guide for the construction of statistical maps intended for cluster detection. The figures also clarify that a number of empirical questions are still awaiting an answer.

A concluding note of caution concerns the risks associated with over-generalization of the results reviewed here. One principle of graphical perception illustrated by Spence and Lewandowsky (1991) holds that questions about what might be 'the best statistical graph' are meaningful only within the context of a particular perceptual task. Thus, while one graph may be superior to another when observers need to, say, identify global trends, the reverse may be true when observers must report single data values. Indeed, Dunn (1988) has shown that symbol maps may have the edge over choropleth schemes when single values must be reported for specific regions. Similarly, Hastie et al. (in press) have shown that categorical color scales support faster recognition of individual values than shaded monochrome maps.

By implication, our survey applies only to the domain of cluster detection, and maps that were shown to be appropriate here may be less adequate to support other perceptual tasks. The importance of this argument is mirrored by the fact that the forthcoming U.S. mortality atlas will include several representations, using monochrome as well as double-ended scales, of each data set (Pickle, Herrmann, Mungiole, & White, in press).

References

Boyle, P., Muir, C. S., & Grundmann, E. (1989). (Eds.), *Cancer mapping*. Berlin: Springer Verlag.

Carswell, C. M., Kinslow, H. S., Pickle, L. W., & Herrmann, D. J. (1995). *Using color to represent magnitude in statistical maps: The case for double-ended scales*. Manuscript submitted for publication. [1]

Cleveland, W. S., & McGill, R. (1983). A color-caused optical illusion on a statistical graph. *The American Statistician, 37*, 101-105.

Cleveland, W. S., & McGill, R. (1984). Graphical perception: Theory, experimentation, and application to the development of graphical methods. *Journal of the American Statistical Association, 79*, 531-554.

Cuff, D. J. (1973). Colour on temperature maps. *The Cartographic Journal, 10*, 17-21.

Dunn, R. (1988). Framed rectangle charts or statistical maps with shading. *The American Statistician, 42*, 123-129.

Gilbert, E. W. (1958). Pioneer maps of health and disease in England. *Geographical Journal, 124*, 172-173.

Hastie, R., Hammerle, O., Kerwin, J., Croner, C. M., & Herrmann, D. J. (in press). Human performance reading statistical maps. *Journal of Experimental Psychology: Applied.* [2]

Lewandowsky, S., & Behrens, J. T. (1995). *Accuracy of cluster detection in mortality maps*. Technical Report, National Center for Health Statistics, Hyattsville, MD. [3]

Lewandowsky, S., Behrens, J. T., Pickle, L. W., Herrmann, D. J., & White, A. (1995). *Perception of clusters on statistical maps: Representing magnitude and reliability*. Manuscript submitted for publication. [4]

Lewandowsky, S., Herrmann, D. J., Behrens, J. T., Li, S.-C., Pickle, L., & Jobe, J. B. (1993). Perception of clusters in statistical maps. *Applied Cognitive Psychology, 7*, 533-551. [5]

Lewandowsky, S., & Spence, I. (1990). The perception of statistical graphs. *Sociological Methods and Research, 18*, 200-242.

Lohse, G. L. (1993). A cognitive model for understanding graphical perception. *Human-Computer Interaction, 8*, 353-388.

MacEachren, A. M., & Brewer, C. A. (1995). *Representing reliability for the NCHS mortality atlas*. Technical Report, National Center for Health Statistics, Hyattsville, MD. [6]

Maher, R. J. (1995a). A history of the influence of statistical maps on public policy. In L. W. Pickle & D. J. Herrmann (Eds.), *Cognitive aspects of statistical mapping* (pp. 13-20). National Center for Health Statistics, Working Paper Series, No. 18.

Maher, R. J. (1995b). The interpretation of statistical maps as a function of the map reader's profession. In L. W. Pickle & D. J. Herrmann (Eds.), *Cognitive aspects of statistical mapping* (pp. 249-274). National Center for Health Statistics, Working Paper Series, No. 18.

Marshall, R. J. (1991). A review of methods for the statistical analysis of spatial patterns of disease. *Journal of the Royal Statistical Society A, 154*, 421-441.

McBurney, C., & Collings, V. (1977). *Introduction to sensation/perception*. Englewood Cliffs, NJ: Prentice-Hall.

Mersey, J. E. (1990). Colour and thematic map design: The role of colour scheme and map complexity in choropleth map communication. *Cartographica, 27*(3), Monograph No. 41, 1-182. [7]

van Oeffelen, M. P., & Vos, P. G. (1982). Configurational effects on the enumeration of dots: Counting by groups. *Memory & Cognition, 10*, 396-404.

Pickle, L. W., Herrmann, D. J., & Wilson, B. F. (1995). *A legendary study of statistical map reading: The cognitive effectiveness of statistical map legends*. Manuscript submitted for publication.

Pickle, L. W., Herrmann, D. J., Mungiole, M., & White, A. A. (in press). Design of the new U.S. Mortality Atlas. In *Proceedings of the International Symposium on Computer Mapping in Epidemiology and Environmental Health*. Tampa, FL, 14 February 1995.

Pickle, L. W., Mason, T. J., Howard, N., Hoover, R., & Fraumeni, J. F., Jr. (1987). *Atlas of U.S. cancer mortality among whites: 1950-1980.* Washington, D.C.: National Institutes of Health.

Sanders, M. S., & McCormick, E. J. (1993). *Human factors in engineering and design.* New York: McGraw-Hill.

Spence, I., & Lewandowsky, S. (1991). Displaying proportions and percentages. *Applied Cognitive Psychology, 5,* 61-77.

Treisman, A. M. (1991). Search, similarity, and integration of features between and within dimensions. *Journal of Experimental Psychology: Human Perception and Performance, 17,* 652-676.

Treisman, A. M., & Schmidt, H. (1982). Illusory conjunction in the perception of objects. *Cognitive Psychology, 14,* 107-141.

Wagner, G. (1989). Recent cancer atlas of the Federal Republic of Germany. In P. Boyle, C. S. Muir, & E. Grundmann (Eds.), *Cancer mapping* (pp. 103-114). Berlin: Springer-Verlag.

Wainer, H. (1992). Understanding graphs and tables. *Educational Researcher, 21,* 14-23.

Wainer, H., & Francolini, C. M. (1980). An empirical inquiry concerning human understanding of two-variable color maps. *American Statistician, 34,* 81-93.

Walter, S. D., & Birnie, S. E. (1991). Mapping mortality and morbidity patterns: An international comparison. *International Journal of Epidemiology, 20,* 678-689.

Acknowledgements

The authors gratefully acknowledge the continued support for this research by the National Center for Health Statistics (NCHS), Bethesda, Maryland. We also thank Linda Pickle and Walter Smith for their helpful comments on earlier drafts of this paper.

VISUAL ANALYSIS OF VERY LARGE MULTIVARIATE DATABASES

Ted Mihalisin, John Timlin, John Schwegler, Ed Gawlinski and Jim Mihalisin, Mihalisin Associates, Inc.
Ted Mihalisin, Dept. of Physics, TempleUniversity, Broad and Montgomery Streets, Philadelphia, PA 19122

Key Words: Multidimensional, Visualization

Abstract

A variety of business, scientific and government endeavors require that one analyze data involving millions of records with many variables including continuous, ordinal and categorical variables. Although some analysis tasks such as sales tracking are well defined, others require exploratory data analysis in order to search for evidence of 2-way, 3-way, ... n-way interactions or conditional correlations which in general may be non-linear. Combinatoric complexity arises naturally due to the multivariate nature of the data or the nature of the analysis itself e.g. combinatoric optimization. We will discuss a system for the visual analysis of these types of databases which is based on U.S. Patent No. 5228119. In the analysis phase the performance of the system is independent of the number of records. New data views are presented in less than one second even for datasets with hundreds of millions of records. Data involving up to 10 dimensions can be graphed and visually analyzed for correlations including highly conditional non-linear correlations. The system will be contrasted with other methods for visually analyzing multivariate data.

Introduction

In 1995 it is hardly necessary to point out the need to analyze very large datasets (databases) containing millions even billions or more datapoints (records). The problem is greatly compounded by the fact that much of the data to be analyzed involves multiple dimensions or variables which in general are a mix of continuous, ordinal and categorical variables. However, even the case of low dimensionality presents non-trivial problems. How do you plot an x-y graph on a monitor containing about one millions pixels when there are a billion x values at which y values are to be represented graphically by the vertical location of a symbol or the vertical extent of a bar or in any fashion for that matter?

The vital need to graph data rather than blindly apply statistical methods is well documented for low dimensional data. The need to graph data is even more pronounced in the case of multi-dimensional data since most common multidimensional or multivariate statistical methods assume variables to be normally distributed. Moreover multidimensional statistical entities such as the correlation matrix make what is tantamount to gross assumptions namely that two variables can be simply classified as correlated or not. But how can one graph multidimensional data? That's the point of this paper and earlier papers[1-9] by the authors stretching back to 1989 when we first published our basic notion which led to U.S. Patent No. 5228119 issued July 13, 1993.

This paper is organized along historical lines.

The first section reviews the methods of graphically analyzing data that existed prior to 1989. The most important method discussed is that of scatterplot matrices[10]. The severe limitations of this technique and for that matter of other techniques which are based on the concept of focusing and linking or on the notion of projections of the data onto subspaces of lower dimensionality are discussed.

The second section introduces the concepts of our TempleMVV system which first appeared in a 1989 paper[1]. In this, the main section, we hope to clarify how we have overcome the difficulties of the pre-1989 methods and have produced a very robust methodology for graphically analyzing very large low dimensional datasets, small but high dimensional datasets and very large high dimensional datasets. Of particular importance is the fact that the method's performance during the visual analysis phase is independent of the size of the dataset.

In the third section we briefly discuss the trellis display method introduced in 1994 by Cleveland et al[11]. This method utilizes one of the concepts that we introduced in 1989 namely the notion of what they refer to as "conditioning", but it differs from TempleMVV in many other vital respects. Their method introduces the ability to look at a mix of 2d and 3d data displays on a trellis of panels, each panel may be thought of as a data display for 2 or 3 variables for a subset of the data selected by constraining or conditioning the remaining variables.

In the fourth section we will discuss what we view as some key differences between the two methods in respect to performance, ease of use, and power. These differences arise from the central MVV notions of extensibility, hierarchical data symbols and hierarchical variable axes, the nature or our totally recursive computational engine, visual perception issues and our rather robust set of data interaction tools.

In the fifth section we briefly discuss some of the advanced capabilities of MVV afforded by the abstraction extensibility of the U.S. Patent No. 5228119. First, we discuss how MVV can be used to plot billions of x-y pairs one million times faster than conventional methods by mapping a low dimensional space to one of higher dimension. Second, how MVV's advanced scaling can be used to visually fit data which may contain high order interactions. Third, how MVV can be used to analyze combinatorial data. And finally fourth, how MVV's three tier architecture allows one to create complex sequences of data views so that analysts need only step through a series of slides and use intuitive but powerful tools.

I. Pre-1989 Methods

Although several methods existed for trying to visually analyze multidimensional data prior to 1989 including glyphs and parallel axes methods as well as scatterplot matrices we cannot review all of them here. Suffice it to say that in our view glyphs[12] and parallel axes[13] approaches are of extremely limited use and are not viable general purpose analysis tools for large multivariate datasets. In this section we briefly point out the severe difficulties present in the scatter plot matrix approach.

Fig. 1 shows schematically a 6 variable scatterplot matrix sans data points. Basically one plots 2d projections of the data points in the 6d space onto all 30 possible choices for the 2 variables (15 if one plots only one side of the diagonal).

In order to see the behavior of the data in 3,4,5 or 6

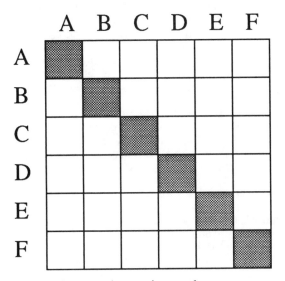

Figure 1 - A scatterplot matrix sans data.

dimensions the user can "brush" the data. Brushing consists of selecting a variable and then constraining it to a subinterval of its values. Brushing can be applied repeatedly and can be applied to more than one variable. As a brush is applied to one or more of the 6 variables all data views in the 30 2d graphs are updated accordingly. Thus if one brushed one variable repeatedly looking at say ten disjoint bins that spanned the variables' range and remembered the corresponding 10 scatterplot matrix views one had in effect seen all possible 3 variable or 3d views of the data involving the brushed variable.

One could then repeat the process for the other 5 variables. A recollection of the say 6x10=60 corresponding scatterplot matrix views provides all 3d views of the data though with some rather confusing redundancy. Similarly if one performed all (6x5)/2=15, two variable brushings each involving say 10x10=100 brushings and remembered the corresponding 1500 scatterplot matrices each involving 4x3=12 2d plots one in effect would see all possible 4d views of the data. 3

variable brushings involve viewing (6x5x4)/(3x2x1) = 20 cases each with say 10^3=1000 brushings for a total of 20,000 views of the remaining three 2d plots in the matrix. Finally the 6x5x4x3/(4x3x2x1) = 15 four variable cases each involving say 10^4 brushings would allow one to see the 6 dimensional behavior of the data.

Therefore rather glaring problems exist for scatterplot matrices when one is trying to view high dimensional data and/or data involving many points or records. These problems and their solution were pointed out in our 1989[1] and subsequent papers[2-9]. First, the scatterplot matrix method requires one to plot all the data at least n(n-1)/2 times, n(n-1) if one wants both sides of the diagonal). Even for small n this leads to an unacceptably long plotting period for even a million points let alone a billion. Second, the need to remember tens of thousands of brushed scatterplot matrices is totally unacceptable even if each were plotted instantly. Humans are great at pattern recognition but not at recalling thousands of patterns. Third, even if one solved the slowness problem and found an individual with an elephantine memory, requiring users to manually step through all possible combinations for brushings is unacceptable. It's a simple example of what we call the "combinatoric burden."

II. Basic TempleMVV

Primitive Cells and Spaces

In the mid 1980's it became clear to us that the only viable way to plot and visually analyze multi-dimensional data was to make all "independent variables" discrete and to allow one or more "dependent variable(s)" to be continuous. Central to the method was the notion of a primitive cell which contained information about the behavior of the dependent variable(s) in a small rectangular n dimensional region (the primitive cell) of the n dimensional independent variable space. That is, all independent variables which are continuous are binned to the finite granularity that the user believes to be appropriate or necessary (MVV itself can be used to assess this granularity). These (now discrete) variables along with intrinsically discrete categorical and ordinal independent variables can be thought of as forming a discrete n dimensional lattice of primitive cells. Although the primitive cells represent n-dimensional objects they are represented graphically as two dimensional objects namely as parallelograms. The angles of the parallelograms are controllable, as are the gaps in each direction between adjacent parallelograms (see Fig. 3). Shown in Fig. 2 is a primitive cell and a variety of symbols that can be placed in the cell namely a "vertical" bar, a "horizontal" bar, a "vertical-horizontal" bar, a circle and a box that can be used to represent the behavior of a dependent variable. That is, the size of the symbol represents the value of some statistic for the dependent variable in the domain of the primitive cell. In addition the color of a cell may be used to indicate the value of a dependent variable statistic. This is particularly useful when tens of thousands of cells are present on the monitor each being rather small. Statistics supported include min, max, mean, sum and standard deviation. Compound symbols for two or more statistics are also shown in Fig. 2. If more than one

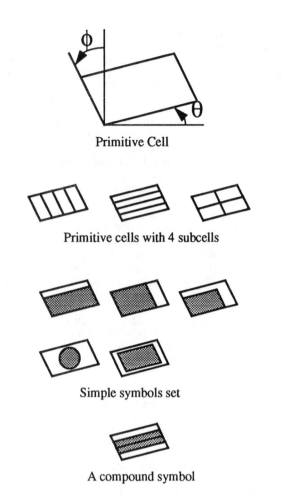

Primitive Cell

Primitive cells with 4 subcells

Simple symbols set

A compound symbol

Figure 2 - MVV cells, subcells, symbols and compound symbols.

dependent variable is of interest the primitive cell can be subdivided into subcells (also shown in Fig. 2) and symbols representing the different statistic-dependent variable pairs can be placed in the subcells. For instance each subcell may refer to the same statistic for say four different dependent variables or four statistics for the same dependent variable or two statistics for one dependent variable and two different statistics for a second dependent variable, etc.

The choice of two dimensional cells and symbols is not accidental. Two dimensional draws are considerably faster than 3d draws. Moreover, we wish to arrange the collection of primitive cells that represent the n dimensional independent variable space in such a way that all primitive cells are completely visible and hence the symbols representing the data are also visible. The geometry engine does this by default whenever the number of pixels required to represent the space is less than the number of pixels in the MVV window. The extensibility of the system to cases where the number of primitive cells exceeds the number of pixels will be discussed shortly. First we need to understand how the geometry engine can arrange the primitive cells in a non-overlapping comprehensible organization. Fig. 3

shows a variety of ways of organizing a simple 6 dimensional collection of primitive cells. For simplicity each of the 6 variables A,B,C,D,E and F has only two bins or values hence there are $2^6 = 64$ primitive cells. Hopefully the way in which each of the representations of Fig. 3 can be extended for the cases where A-E have arbitrary and in general unequal number of bins or values is obvious. Consider Fig. 3a which shows the simplest representation namely one in which all primitive cells are lined up in a horizontal fashion with variable A nested inside B, B inside C etc. Here we say that there is one hierarchical axis namely a 6-fold horizontal hierarchical axis. Fig. 3b shows a related cell arrangement in which variables A and B are running in directions which are not parallel nor are they parallel to the remaining variables' direction i.e. C,D,E and F which are all horizontal. MVV allows users to specify the directions of each variable axis via an angle of departure from either the vertical or the horizontal. Note that although the directions of A and B differ from one another and also from the direction of C, D, E and F nonetheless A is still in a sense nested inside B which is nested in C etc. Another way of pointing this out is to say that the graphical metric for a step of one in A i.e. from A_i to A_{i+1} is smaller than a step of one for B from B_j to B_{j+1} etc. Fig. 3c shows a two hierarchical axis MVV arrangement of primitive cells with A, B and C along the horizontal and D, E and F along the vertical. Finally Fig. 3d shows a two hierarchical axis nesting but with six distinct directions for A - F. MVV allows the user the freedom to create as many hierarchical axes as he or she desires and to control the directions of all variables and the gaps between primitive cells (and as we will see later the gaps between hierarchical cells as well), as well as the angles of the parallelogram primitive cells themselves, the type of symbols used and the number of sub-cells corresponding to the number of dependent variables. Even this basic type of TempleMVV graph is extremely powerful.

Extensibility: Color Maps and Hierarchical Cells and Symbols

In many cases one might have "too many" primitive cells. That is, the size of the resulting primitive cells might be too small to draw meaningful symbols. For example if a primitive cell is a rectangle 5 pixels by 5 pixels the allowed heights of say vertical bars would only be 0, 1, 2, 3, 4 or 5. Hence precise determination of a dependent variable's statistic would be impossible and moreover it would be difficult to read the resulting myriad of short vertical bar charts. One way to extend the capability of MVV is to represent the value of a dependent variable's statistic by a color map filling the entire cell. Although color is a more qualitative representation of a number than is physical size it nonetheless can be extremely useful for a variety of analysis tasks.

A more fundamental way of extending MVV is by defining hierarchical cells. Hierarchical cells represent "roll-ups" of smaller cells. In Fig. 4 we show one example of a hierarchical cell set that results from rolling up one or more variables of Fig. 3c. The basic TempleMVV system allows one to view not only the

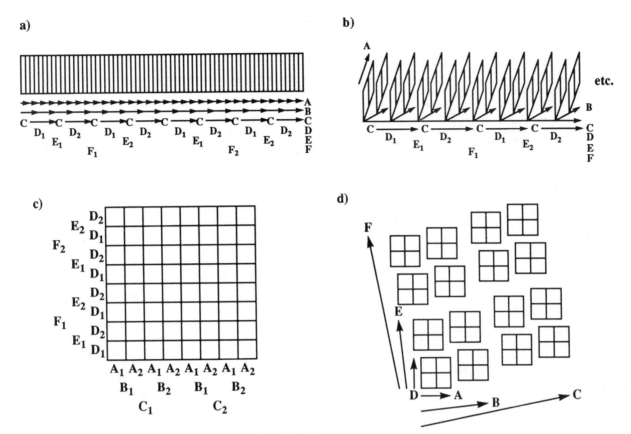

a)

b)

etc.

c)

d)

Figure 3 - For different ways to lay-out a 6 dimensional space.

primitive symbols and cells (if they are large enough) but as many hierarchical symbols as one wishes. This allows one to look at information of different dimensionality at the same time. At this point it might be instructive to look at a concrete example of an MVV graph. Fig. 5 shows a simple MVV graph of car sales by year, model and region. Here the number of cars sold determines the height of each vertical bar. The black bars are the primitive symbols. There is one black bar for each year-model-region triplet i.e. a total of 3x4x4=48. The gray bars represent roll-ups over both year and region using the sum statistic and hence there are only 4 gray bars one for each model. The heights of each gray bar represent the total sales for a given model irrespective of (summed over) year and region. Note that the scales for black and for gray bars are not indicated on the graph a point that will be discussed later.

In Fig. 6 we show the same data but the variables year, region and model all run at non-zero angles with respect to the vertical (the direction of the symbol) and the 3 years are "color" coded white, gray and black. This minor variation reveals a very important point. In Fig. 5 one first thinks of the black bars as an organized set of vertical bars representing a time series over the years 92, 93 and 94. There are 16 such series for the 4x4 = 16 region-model pairs. Once can also think of the black bars in Fig. 5 as 4 sets of 12 bars each set being a graph of sales versus time and region for a given model. One can of course think of these 12 bars as 4 time series one

for each region and focus on how the time series differs from region to region. But is more difficult to think in terms of, or to visualize, how sales depend on region in a given year for a given model or how sales depend on model for a given year and region when viewing Fig. 5. In the former case it's because the bars and the variable direction are both vertical hence the variation is longitudinal rather than transverse. In the latter case its because year is nested inside of model and its difficult for the eye to skip over the years 93 and 94 to track how the black bars' heights are varying from 92 Model A to 92 Model B to 92 Model C and finally to 92 Model D. On the other hand the angles and the color coding in Fig. 6 make it far easier to see the full year, region and model dependence of sales rather than just the year dependence and how the year dependence evolves with model and region (as was the case for Fig.5). The eye can follow how the white vertical bars (92) vary with model and region as well as how the 3 contiguous bars vary with year (white 92, gray 93 and black 94). This is an extremely important point since without angles one would normally have to show n! nesting orders for a one hierarchical axis display or (n-m)!(m)! for a two hierarchical axis case when m independent variables are along say the horizontal and n-m are along the vertical.

In Fig. 7 we show an extension of Fig. 5 that adds two more variables namely incentive plan (3 values i.e. A, B and C) and buyer income (3 brackets). Now there are 5 independent variables. The black bars show sales for each year, model, region, buyer income bracket and

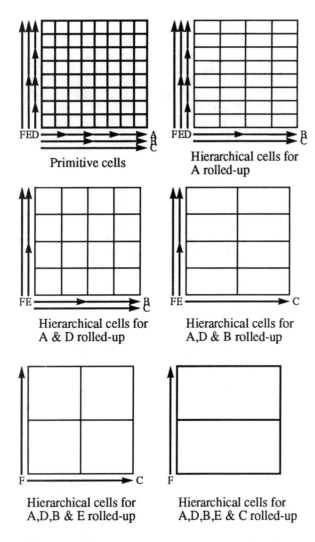

Primitive cells

Hierarchical cells for A rolled-up

Hierarchical cells for A & D rolled-up

Hierarchical cells for A,D & B rolled-up

Hierarchical cells for A,D,B & E rolled-up

Hierarchical cells for A,D,B,E & C rolled-up

Figure 4 - Primitive cells and 5 types of hierarchical cells for a 6d space.

incentive plan quintuplet for a total of 3x4x4x3x3 = 432 black bars.

Again the gray bars show roll-ups over year and region but these now depend on not only model but also incentive plan and buyer income for a total of 4x3x3 = 36 gray bars.

Tools

The importance of fast interactive tools, status indicators and simple controls can never be overstated. Without tools the multidimensional graph becomes a static, monolithic presentation which often raises more questions than it answers. It is essential that the user of a multidimensional graphing system be able to make immediate and direct changes to and on the graphs which are guided by the information presented in the graphs. These tools should allow the user to explore the represented data as generally as possible. This can be accomplished with a small set of tools whose functions are independent of one another. Each tool can be applied

to the graph irrespective of the number or type of tool(s) previously applied. We call such a toolkit "multiplicative".

The tools in this multiplicative tool kit should apply to both the independent variable space and the dependent variable symbols. In TempleMVV this provides a natural division constructed to provide simplicity in the status and control elements and therefore a clear representation of the current graph's state. Tools which operate on the independent variable space should allow the user to zoom, subset, granulate, and reorder the values of the independent variables. The tools that operate on the dependent variable symbols should allow the user to select symbols for plotting based on user preference, desired ranges and other dependent variable/ symbol related criteria. The driving force for the selection and application of a tool should always be the current graph — i.e. the data, not just pre-conceived queries.

The current collection of tools for both the independent variables and the dependent variables in TempleMVV are applicable only to one variable at a time. This greatly simplifies the number of steps required to effect the most common and most powerful multidimensional operations. The response of TempleMVV to the application of a tool is immediate allowing the newly produced graph to give the user valuable feed-back. The currently available independent variable tools in TempleMVV are:

- Zoom In/Zoom Out
- Animate (a collection of sequential zooms)
- Subset/Restore Set
- Decimate/Undecimate
- Coarser Grain/Finest Grain
- Sort Zoomed/Original Order
- Visual Reorder/Original Order
- Rename Value

The zoom tool allows for quick selection of interesting subspaces and provides a selection mechanism for the animate and sort zoomed tools. In effect these three tools allow the user to search for trends in any subspace and to sort values of an independent variable based on the behavior of the dependent variable in any contiguous subspace. The subset and decimate tools allow the user to select subranges of variable values and recomputes all summaries dependent on these changes. The coarser grain tool allows the user to combine bins together for a coarser view. The visual reorder and rename value tools let the user select any desired value order and rename values as desired.

The currently available dependent variables tools are:

- Symbols On/Symbols Off
- Select Bins/Reset Bins
- Normalize On/Normalize Off

The symbols on/off tool lets the user select which

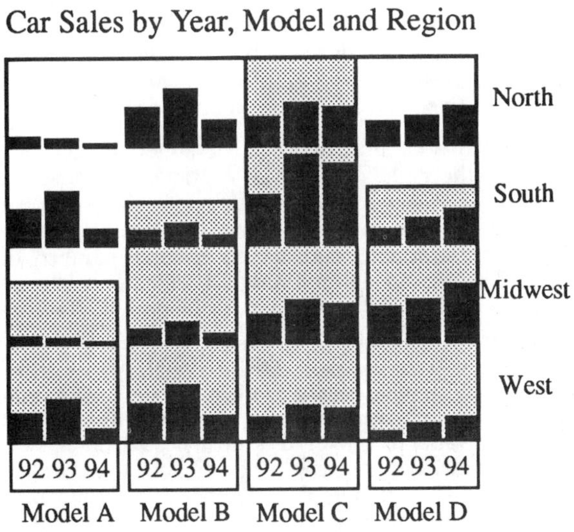

Car Sales by Year, Model and Region

North

South

Midwest

West

| 92 93 94 | 92 93 94 | 92 93 94 | 92 93 94 |

Model A   Model B   Model C   Model D

Region

Model

Year

Figure 5 - A "zero angle" 4d plot (3 independent variables.

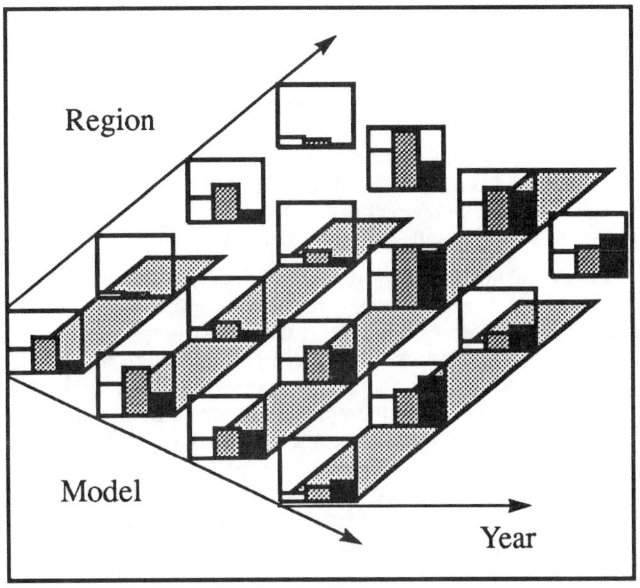

Figure 6 - A "non-zero angle" plot.

Car Sales by Year, Model, Region, Income and Incentive

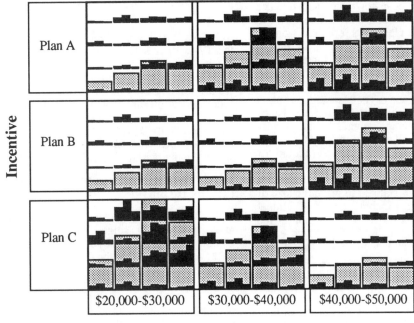

Incentive

Plan A

Plan B

Plan C

| $20,000-$30,000 | $30,000-$40,000 | $40,000-$50,000 |

Buyer's Income

Figure 7 - "Drilling" the data of figures 5 and 6 two independent variables deeper (to 5 independent variables).

summary level hierarchical symbols (roll-ups) are to be displayed. The select bins tool allows the user to select ranges for each summary level, suppressing symbols outside the specified ranges and graphing the ones contained thus producing a multidimensional contour plot of the data. The normalize tool allows the user to separate the scales of all symbols at the chosen level of summary. This tool is useful for displaying pattern shifts and changes which generally would be indiscernible due to large variations in scale (see Section V).

The list of tools currently available is by no means an exhaustive one, however, the use of these tools in conjunction with one another produces a large number of possible display states at the touch of a button. The display states the user actually selects are guided by the current data view and the analysis interests of the user. This is critical since as the dimensionality increases the number of possible views increases astronomically. Any multidimensional system which relies on a full scan or "grand tour" of the whole space or a static picture of the space for analysis will quickly frustrate the analyst. The truth is that multidimensional data usually contains many interesting results and not just a single message. A good multidimensional visualization system must contain real time interactive tools to allow the analyst to pursue any line of exploration.

Statistics and Robustness

As stated earlier a variety of statistics may be used to drive the size and/or color of a primitive or hierarchical symbol. Some statistics, however, are less robust than others in that they require a particular distribution function or class of distribution functions in order to be meaningful. In Fig. 8 we show the mean income± the standard deviation of the mean for males and females with a variety of education levels and ages. The white bars represent men and the black bars represent women. The top of each vertical bar is at the mean plus a standard deviation of the mean income for a man or woman of a given education level and age while the bottom of each vertical bar is at the mean minus a standard deviation of the mean. The standard deviation is computed assuming a normal distribution and hence is not robust. Fig. 8 may suggest that gender discrimination occurs for all education levels and age groups. MVV provides a completely robust method for analyzing this data. Namely, one lets data point or record count be the dependent rule, i.e. here the number of people, and uses the sum statistic to drive the size of the symbol and introduces income as an additional independent variable. Fig. 9 shows such a graph for the same data used to create Fig. 8. In Fig. 9 we see a striking effect for men and women at the higher education and age levels. In particular the distribution versus income for men at high education and age levels develops a secondary peak at high incomes while the distribution versus income for women doesn't. Thus it would appear that at least as of the 1990 census women as a group had not been promoted to the highest positions or perhaps had not yet penetrated the highest paid professions. A further breakdown of the data by type of occupation would clearly be useful. It should be noted that MVV's fast computational engine is 100%

recursive so that certain statistics such as the median and mode etc. are not used. We would maintain that visualizing the entire distribution as in Fig. 9 is in fact superior to box-plot type graphics[14] which require the use of statistics that cannot be computed recursively.

In Figure 8 note that two sets of slider widgets are shown to the right of the MVV graph. The first set (column of 3 widgets) is labeled "Independent Variables". These widgets show the name of each independent variable and its range and number of values. The second set (column of 3 widgets) is labeled "Dependent Variable", "Mean +- of Income I". The top most Dependent Variable widget refers to the primitive symbols (no roll-ups) and hence depends on all three variables S, E and A (sex, education and age). The middle dependent variable refers to the roll-up of the dependent variable I over the most nested or "fastest running" variable namely sex. Hence this widget represents the 1st hierarchical symbol namely mean +- of income rolled-up over gender and thus depends only on E and A. The X through the vertical bar symbol on the widget indicates that these symbols are turned off. The bottom Dependent Variable widget corresponds to the 2nd hierarchical symbol namely a roll-up over both sex and education. The corresponding symbols (turned-off in Figure 8) would show mean income +- a standard deviation of the mean versus one variable namely age. All of the tools discussed earlier can be invoked using the mouse on these widgets or in the case of IV tools on the hierarchical axes of the graph or on a list box named the item manager which lists the values for each independent variable. The item manager is particularly useful when one or more variables have a very large number of values and of course for categorical variables generally since categorical variables have no natural order.

In Figure 9 only the IV widget set is shown. Note that the IV widget set indicates that we have subset the income variable so that we are looking at the first 29 of 62 values which run from -10,000 to +300,000 and that we are looking only at the top 2 of 6 education levels. Since education runs from 0 to 18 years, and here equal width bins have been chosen, each of the 6 bins corresponds to 3 years of education hence the two education levels are 12 to 15 years and 15 to 18 years as shown on the hierarchical axis.

III. Arranging and Grouping Graphs on a Page (Trellis Displays)

Applications have allowed multiple symbols types, axes and graphs on a single page since the origins of computer graphics. The primary problem with these packages has been that the construction of collections of graphs with multiple common axes and/or aligned axes has been tedious to say the least. Recently, there has been an explosion of applications which allow for automated arrangement of symbols, axes and graphs on the page. Some of these packages are better than others at achieving the desired end with minimal user interaction. The Trellis display system[10] appears to be an effective programming paradigm for displaying collections of graphs typically used in statistics. The basic premise behind these packages is that existing 1d,

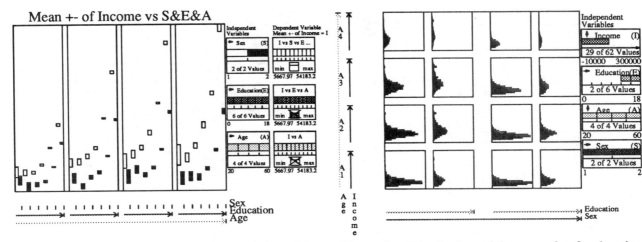

FIgure 8 - Mean income +- the standard deviation of the mean versus gender, education and age.

Figure 9 - Distribution of incomes for 2 education levels, 4 age groups and both genders.

2d and 3d graph constructions represent a modular technology which needs an object oriented system of geometry management. The beauty of this philosophy is that the graphing engines and the geometry management systems remain essentially independent of one another.

This type of multiple graph geometry management is currently used in many diverse applications in the industry. The common thread in these applications is the need for related graphs to share a common display, common scales, common horizontal and/or vertical positions, etc. and these packages provide a graph wizard or quick graphing features for no think graph production. The primary usage of the multi-graph systems is in report writing, status displays of real time data and multi-frame plotting. These are simple, straightforward presentations of related graphs with minimal interactive features.

IV. Arranging Graphs vs Exploratory Data Analysis

Although it is tempting to expand the domain of the graph arranging packages from that of presenting collections of graphs to the domain of exploratory data analysis, there are many reasons why this philosophy won't provide an effective exploratory visual analysis system. Key among them is the basic design of these packages. An effective exploratory visual system requires an enormous amount of crosstalk between the graph constructor and the geometry management system. Graph arranging packages are not designed to handle this crosstalk. In a trellis-like collection of graphs or "panels" the positions of the panels and their relationship to the constrained variables is just as vital as the contents. The vast majority of possible panel arrangements do not provide multidimensional insights.

An extensible, interactive visual system must have as its atomic module an object which is both intrinsically simple and one which can be hierarchically placed in an extensible manner such that interactive manipulations produce sensible, readable composite structures. A single graphic symbol, not an entire graph is the solution

and the reasons are not hard to understand. Using the graph as an atomic module introduces a number of problems, namely: 1) a bias of the display of the data which unnecessarily introduces combinatoric problems which unbiased symbols avoid, 2) a lack of visual extensibility i.e., a graph as a module cannot be easily extended to a simple summary for more compact viewing, whereas, a simple collection of symbols can be summarized into a single summary symbol, 3) longitudinal variation - it is difficult to construct and arrange graphs so that intergraph comparisons produce transverse symbolic variation, 4) the pixel overhead of the graph and its labels, axes and frame make viewing data of higher depths (dimensions) unmanageable and 5) discontinuous viewing - simple graphs such as scatterplots allow for the (x, y) locations to vary from point to point and frame to frame making comparison between frames difficult if not impossible, a symbol as an atomic modular construct provides a uniform positional definition across the entire view.

V. Advanced TempleMVV

TempleMVV has a very extensible structure that allows it to handle a very wide range of analysis and data types. For example using what we call modulo variables and hyphenated variables TempleMVV can turn a low dimensional dataset into a higher dimensional one and vice-versa if it serves a useful purpose. For example if one has a time series involving a dependent variable evaluated at one billion equally spaced times, TempleMVV can be used to create 3 variables or dimensions each with 10^3 values with 1 hierarchical axis. Since there are only 1000 or so horizontal pixels only the 1000 2nd hierarchical symbols can be shown. Each of these symbols correspond to a roll-up over 10^3 1st hierarchical cells each of which corresponds to a roll-up over 10^3 primitive cells. Hence each 2nd hierarchical symbol corresponds to a roll-up of 10^6 primitive cells. One uses the min/max statistic to drive the location of the bottom/top of a vertical line segment for each 2nd hierarchical symbol. This produces a graphic identical to a polyline routine except one

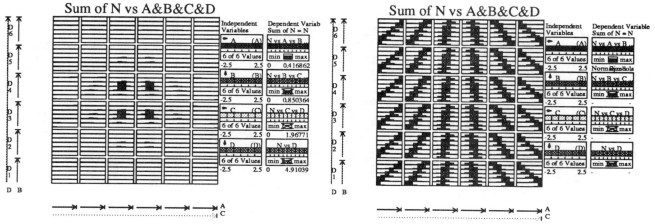

Figure 10a - Number or count versus 4 variables with all vertical bars or the same scale.

Figure 10b - same data as in 10a but each bar chart N vs A has its own scale.

million times faster. One can then use all of the tools described earlier to explore the dataset at subsecond speed.

Another very important way in which MVV is extensible is in the realm of symbol scaling. Consider first Figure 10a which shows an MVV graph where the dependent variable is count or number and the rule is sum with 4 independent variables A, B, C and D each with 6 bins. This four-variate distribution function is basically normal but with some interactions. In Figure 10a we show only the primitive symbols or vertical bars. All vertical bars are on the same graphical scale. As a result only bars near the centers of the ranges of all four variables have appreciable height. Hence it is very difficult to decide on an appropriate model for the data. But Figure 10a is just one of a very large number of ways to scale or normalize the data.

Consider next Figure 10b. Think of Figure 10b as a collection of 6^3 histograms of N vs. the fastest running variable A. In Figure 10b each histogram has its own scale which is determined to be equal to the value of the largest of the 6 vertical bars that correspond to N vs. A for fixed values of B, C and D. That is each of the small bar charts of Figure 10a has been "scaled-up" so that it is easily visible.

When one looks at Figure 10b one should note that for fixed C and D values the histograms shift with varying B. This indicates an A-B interaction. Similarly the N(A) histograms shift as you vary C for fixed B and D indicating another 2 way interaction namely A-C. But the N(A) histograms do not shift as you vary D holding B and C fixed i.e. there is no A-D interaction. Next focus your attention on the pattern formed by the collection of 6 bar charts obtained by varying A and B for fixed C and D. Note that this pattern changes with C for fixed D hence there is a 3 way interaction of A-B-C. But this pattern does not change as you vary D for fixed C that is; no 3-way interaction involving A-B-D exists. Finally the pattern formed by the 36 bar charts of form N(A) obtained by letting B and C vary does not change from one D value to the next this means that a 4 way A-B-C-D type interaction is not present. By selecting each of the 6 variables as the fastest running and using the procedure just outlined one can determine the highest order of interaction for each

variable. TempleMVV allows an expert user to create such a macro so that analysts less well versed in TempleMVV can simply step through the relevant data views and use their pattern recognition skills. TempleMVV in fact has a 3 tier architecture one for experts called TempleMVV "Developer" one for analysts ("Analyzer") and one for less technical staff called "Viewer" which is basically a slide show created by either an expert or an analyst.

One final note on advanced TempleMVV capabilities. TempleMVV has a hierarchical object oriented structure that allows it to create new metaspaces which themselves contain MVV graphs. These spaces involve meta-variables which dictate the selection and/or nesting order of variables to be included in the metaview. These variables are intrinsically combinatoric in nature and hence can either describe a meta-space of MVV graphs in which each primitive cell contains an entire MVV graph or simply one number or several numbers (dependent variables) for a less abstract combinatoric problem. In Figure 11a we show the case of 2 metavariables one along the horizontal and one along the vertical. Each metavariable runs over all 8 ways of selecting 0, 1, 2 or 3 distinguishable objects from a group of 3 objects. Of course there is nothing sacred about this meta-dimensionality. Figure 11b shows the case of 3 metavariables. Each metavariable in Figure 11b runs over the 4 ways of selecting 0, 1, or 2 distinguishable objects from a group of 2 objects. If A-F represent IV's in an MVV then each cell in Figure 11a and Figure 11b represents an MVV graph of the dimensionality indicated by the number in the cell. If A-F are simply objects then each cell represents a drawing of from 0 to 6 distinguishable objects. Figure 11a partitions the six objects into 2 groups of 3 while Figure 11b partitions them into 3 groups of 2. MVV allows for arbitrary groupings. Since one or more figures of merit can be plotted in each cell (or its subcell) this type of representation is clearly of relevance to fields involving for instance combinatoric chemistry.

a)   **b)**

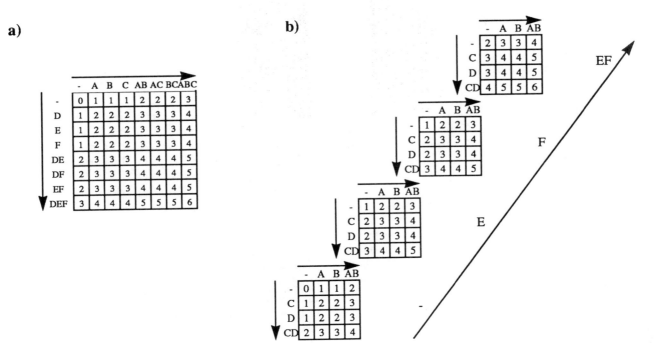

Figure 11 - a) A combinatoric space or metaspace involving two combinatoric variables or metavariables.
b) A combinatoric space involving three combinatoric variables or metavariables.

References

1. Mihalisin, T., Gawlinski, E., Timlin, J. and Schwegler, J., Scientific Computing and Automation Vol. 6, No. 1, Oct. 1989, pp.15-20.

2. Mihalisin, T., Suntech J., Vol. 3, No. 1, winter 1990, pp.25-31.

3. Mihalisin, T., Gawlinski, E., Timlin, J., and Schwegler, J., Proc. IEEE Conf. Visualization, San Francisco, Oct. 1990, pp. 255-262.

4. Mihalisin, T., Timlin, J., and Schwegler, J., IEEE Computer Graphics and Applications, Vol. 11, No. 3, May 1991, pp. 28-35.

5. Mihalisin, T., Timlin, J., and Schwegler, J., Proc. IEEE Conf. Visualization, San Diego, Oct 22-25, 1991, pp. 171-178.

6. Mihalisin, T., Schwegler, J., and Timlin, J., Proc. 24th Symposium on the Interface, College Station, March 18-21, 1992, ed. H. Joseph Newton, Interface Foundation of America, Fairfax Station, VA, pp. 141-149.

7. Mihalisin, T., Schwegler, J., and Timlin, J., 1992 Proc. of the Section on Statistical Graphics, Boston, Aug. 9-13, 1992, American Statistical Association, Alexandria, VA, pp. 69-74.

8. Mihalisin, T., Schwegler, J., Gawlinski, E., and Timlin, J., 1993 Proc. of the Section on Statistical Graphics, San Francisco Aug. 8-12, 1993, American Statistical Association, Alexandria, VA pp. 83-88.

9. Mihalisin, T., Schwegler, J., Gawlinski, E., Timlin, J., and Mihalisin, J., Proc. 26th Symposium on the Interface, Research Triangle Park, June 15-18, 1994, ed.s John Sall and Ann Lehman, Interface Foundation of America, Fairfax Station, VA pp. 426-430.

10. Becker, R.A., Cleveland, W.S., and Wilks, A.R., in "Dynamic Graphics for Statistics", eds. W.S. Cleveland and M.E. McGill, Wadworth and Brooks/Cole, Belmont, CA, 1988, pp1-12.

11. Becker, R.A., Cleveland, W.S., Ming-Jen, S., (previous paper in these proceedings).

12. Bergeron, R.D., and Grinstein, G., Proc. Eurographics 89, North Holland, Amsterdam, 1989, pp. 393-399.

13. Inselberg, A., and Dimsdale, B., Proc. IEEE Conf. on Visualization, San Francisco, Oct. 1990, pp. 361-375.

14. Tukey, J.W., Exploratory Data Analysis, Addison-Wesley, Reading, MA, 1997 chaps 2-4.

INCORPORATING DENSITY ESTIMATION INTO OTHER EXPLORATORY TOOLS

David W. Scott, Rice University
Department of Statistics, MS 138, POB 1892, Houston, TX 77251-1892

Key Words: Density Estimation, Exploratory Data Analysis, Grand Tour, Scatter Diagrams, Parallel Coordinates, Averaged Shifted Histogram

Abstract

Preliminary understanding of a new data set is routinely accomplished with graphical tools, such as those popularized originally by EDA. A number of more recent ideas for multivariate data analysis have emerged and some are available in software packages or shareware such as XGobi. In this talk, we illustrate how many of the point-oriented techniques can be supplemented by incorporating nonparametric density estimates. Examples from the grand tour to parallel coordinates to clustering will be presented. Potential advantages include visual simplicity, recognition of unusual structure, and handling an additional dimension.

1. Introduction

Research in statistical graphics is driven about equally by abstract ideas and by hardware innovations. This paper focuses on somewhat abstract ideas for which the latter is becoming relevant. Computers, software systems, and practicing statisticians are being challenged and stressed in the new data environment. Here, we discuss software solutions that will improve the efficiency of the human client and offer new capabilities.

The evolution of the central components of a graphics workstation has not been even. Screen resolution most rapidly "saturated," reaching the quite useful 1280 by 1024 pixel mapping available on most machines today. Higher resolutions are available, but there does not seem to be much pressure in the marketplace for such improvements, compared to the pressure when low resolution screens dominated the PC market. CPU speed has had an impressive but fairly constant rate of improvement.

After lagging behind for years, there has been substantial and rapid progress in the size of computer hard disks and CPU memory available. A Unix workstation can reasonably be outfitted with 10GB disk, 512MB CPU, and the serial power of a Cray. The existence of this class of workstation has encouraged the growth of on-line and very large data sets.

Initial experience by workers trying to understand these *Massive Data Sets* using graphical tools has been surprisingly negative. The hardware bottleneck is caused primarily by the slow graphical I/O, as well as by the slow disk I/O.

Thus we are in the situation of having sufficient data to warrant extensive graphical exploration, but hardware and software which are not (yet) up to the task. Graphical I/O and advanced visualization have not been top priorities for most manufacturers, and progress has only been steady, not rapid. The challenges of Massive Data Sets require simultaneous improvements in graphics I/O and screen resolution.

It is doubtful that computer monitors will ever have significantly more than several millions of pixels, and direct exploration of a billion data points will require compromise and innovation. One compromise is the selection of subsets of data, either randomly or by choosing chunks of time. One innovation is to build and design nonparametric tools into the software packages that currently support EDA. This paper describes how *Multivariate Density Estimation* (Scott, 1992) is well-suited for this task. Some items on our wish list of new tools cannot be accomplished with current generation hardware. Some will require multiple processors with tightly synchronized output to frame buffers for smooth animation viewing.

2. Existing Tools

Interactive tools for exploring multivariate data are widely available, but none is more attractive or complete than XGobi (Swayne, Cook, and Buja, 1991). This one well-designed program incorporates a number of features and techniques: jittered dot plots for univariate data; pairwise scatter plots for bivariate data; rotating scatter plots for trivariate data; and the grand tour (Asimov, 1985; Buja and Asimov, 1986) for higher dimensional data. Under the grand tour option, automated search for 2-D structure in the data can be accomplished by projection pursuit (Friedman and Tukey, 1974; Cook, Buja, and Cabrera, 1993). Exploration is aided by identification tools such as brushing, grouping, subsetting, and identification. Subsets of points can be highlighted through color, size, or shape of glyph (Cleveland, 1993).

Of course, it is easy to suggest enhancements to

the basic XGobi features. For example, the univariate dot plots are only available one at a time — having side-by-side dot plots would be a nice addition. Also missing is the scatterplot matrix option. A scatterplot matrix of selected variables might also allow the user to select a single pairwise scatter diagram by pointing and clicking with the mouse (an easier interaction sequence than the current method of selecting variables manually or stepping through all pairs). More challenging enhancements might include trivariate projection pursuit options to feed into the rotation tool.

XGobi has recently added the ability to draw lines in addition to points. Thus elementary graphs or maps can be included. One alternative to scatter diagrams that uses lines is the parallel coordinate plot (Wegman, 1990)—a tool that might usefully be included.

A very advanced feature allows multiple XGobi windows to communicate with each other, and synchronize their actions. This feature facilitates general types of "linking" such as brushing in one window while viewing other variables in a second window (or even in three windows).

3. Smoothing Scatter Diagrams

XGobi provides a useful model for an exploratory tool. The idea of incorporating density smoothing is not new. In fact, Hurley and Buja (1990) demonstrate the ability to display the averaged shifted histogram (Scott, 1985) in XGobi. The ASH bins the data and applies a local convolution smoothing. The output is a matrix of the form $(t_i, f(t_i))$ where i ranges over the number of bins and $\{t_i\}$ are points at the bin centers. The ASH points (or lines) can be displayed. This feature is not implemented in the current XGobi release.

A different use of density smoothing was demonstrated by Stuetzle (1987) who suggested linking to a histogram, in which every data point is a "building block" stacked in bins that can be brushed. For our purposes, the smoothness of the ASH is relevant, as we propose to view sequences of ASHs based on grand tour projections. The ASH provides an essentially continuous view of the data as they are being rotated. As noted by Hurley and Buja, even small changes in the 1-D projection angle can result in distracting jumps in the histogram.

The term "scatterplot smoothing" can be applied equally to regression data (Cleveland, 1993) and non-regression data (Scott, 1991). The data we have in mind do not have a response variable, and hence the latter (but less common) meaning is assumed.

We have argued that the scatter diagram points

to the density plot, especially as $n \rightarrow \infty$ (Scott and Thompson, 1983). Scott (1992) has discussed in detail the use of the ASH for visualizing data in 1, 2, 3, and more dimensions. We have found the bivariate ASH in particular an extremely effective tool for visualizing data coming out of the grand tour. The computational horsepower of today's machines is pushed to the limit while attempting to display an evolving bivariate ASH in real time. However, we are convinced that the contour surfaces of the 3-D ASH will eventually provide an even more compelling view of data coming out of the grand tour. However, the Silicon Graphics (SGI) computer we used for this work (Model 310GTX with hardware transparency) can barely handle one frame at the bin resolution we desire. Thus our 3-D examples are all oversmoothed to permit some semblance of the animation. We call such a sequence the *density grand tour*.

4. Visualizing Multivariate Densities

Let $\hat{f}(\mathbf{x})$ denote the ASH density estimate, which is really an array of values over a mesh. Depending on the dimension, we can view $\hat{f}(\mathbf{x})$ directly (graph or perspective plot) or indirectly through its contours (2 or 3 dimensions). We identify contours of $\hat{f}(\mathbf{x})$ as α-shells denoted by S_α; specifically, S_α is the set of points $\mathbf{x} \in \Re^d$ satisfying

$$S_\alpha = \{\mathbf{x} \in \Re^d \mid \hat{f}(\mathbf{x}) = \alpha\hat{f}(\mathbf{x}_m)\}$$

where $\alpha \in (0,1)$ is a real number and $\mathbf{x}_m$ is the global mode

$$\mathbf{x}_m = \arg \max_{\mathbf{x}} \hat{f}(\mathbf{x}).$$

For example, if $\hat{f}$ is a trivariate normal density, then an α-shell is an ellipsoidal solid (surface). As α varies, only the size and not the shape of the ellipsoid changes.

For $\mathbf{x} \in \Re^1$ or $\Re^2$, we graph the function $\hat{f}(\mathbf{x})$ itself, rather than its contours. However, in the bivariate case, if the ASH function surface is color-encoded by height, then viewing the density directly from above reveals the α-contours; see Scott and Salch (1995).

5. Examples of the Density Grand Tour

We implemented a prototype of the DGT in Splus, using the Geomview software on an SGI for visualization. We re-coded the ordinary grand tour (GT) so that we could do scatter diagrams and ASHs on the same projection views. Our GT follows the full-dimensional implementation of Wegman, Carr and Luo (1993), so that we could examine 3-D ASHs as well. Note that Wegman can display all the grand tour variables with parallel coordinates.

5.1. One-Dimensional Density Grand Tour: Iris Data

Figure 1 displays a subset of the 1-D DGT of the iris data at three different "times." The data have been subjected to a one-time jittering in the *vertical* direction, so that points move only left or right during the GT. XGobi displays dot plots vertically, but we choose a horizontal display to parallel the ASH x-axis. The corresponding DGT ASHs are shown to the right of the data point frames. Discerning the number of modes and clusters in the jittered dot plot is facilitated by examination of the corresponding ASH. The number of groups in the data is not always as clear as these three frames suggest. Clearly the jittered dot plot and ASH should (and can) be displayed simultaneously if n is not too large. For large n, only the ASH presentation is feasible.

A one-dimensional scatter diagram does not hold much promise, and the expectation for the ASH view may not be much greater. However, there is a certain fascination watching the ASH undulate while simultaneously watching points sliding back and forth across the screen.

Of course, the search for multivariate *structure* requires multivariate views. But the one-dimensional views allow for discovery of very simple grouping structure. Unless the multivariate structure in each group is likely to be substantially similar, subsetting at this time is probably a prudent strategy at some stage of the exploration process.

5.2. Two-Dimensional Density Grand Tour: PRIM7 Data

The PRIM7 data have much structure, as indicated in Figure 2. These data were used by Friedman and Tukey (1974) to illustrate their work on projection pursuit. The view at $t = 64$ shows most of the data piled into one clump with two "arms." A second view at $t = 2$ very effectively shows three clusters falling roughly in a triangular shape. All four frames are at the same viewing angle, but the scaling for the points and ASHs is not exactly the same.

Friedman and Tukey found a Z-shaped structure in these data. Jee (1987) used a different PP criterion and obtained a more compelling view of a triangle with points clustered in the vertices even more pronounced than in Figure 3 at $t = 2$. Cook, Buja, and Cabrera (1993) found pairs of whiskers at each vertex.

The eye is very capable of discerning discrete structure or grouping, even if it is changing rapidly.

The eye is much less successful as discriminating based on density or concentration of data. The height (or color) of the ASH is invaluable for this secondary goal; compare the ASH and scatterplots again at $t = 64$.

Notice that the ASH estimates tend to be somewhat oversmoothed in this figure. Users should have the smoothing parameter at their control, together with a history replay mechanism to dig deeper into recent views for golden nuggets of structure and information.

5.3. Three-Dimensional Density Grand Tour With Brushing: Iris Data

One frame of the data in a GT is shown in the upper left frame of Figure 3. By brushing on the species indicator variable, three ASH contours (all at the same α-level) are shown, one for each species; see the upper right frame of Figure 3. This view is in contrast to the corresponding view (same projection) without the labeling information; see the lower left frame in Figure 3. The first ASH shows the *Setosa* variety quite clearly, but at a higher α level, the exploratory view displays 4, not 3 modes; see the lower right frame in Figure 3. The weight of evidence in favor of the 4th mode can be determined to be slight, as views of the data in nearby projections show the presence of a long and narrow ridge in the bigger contour, that sometimes shows 1, 2, or 3 modes, but never at much "depth." By depth, we mean the height of the density halfway between modes. Not much depth is observed at any projection.

6. Controlling the Density Grand Tour

While a number of design issues go into the construction of an ASH, the basic controlling knob goes to the smoothing parameter $m = h/\delta$; here δ is the binning resolution and h is the usual kernel bandwidth. Our examples in the previous section clearly indicate that useful views of the data are provided even when h is far from "optimal." Nevertheless, a user should have the ability to control h in real time.

Color is a relatively crude marker (as anyone who has tried to use color to brush scatterplots with large n and several groups knows). We have found it useful in two different ways. The first is to differentiate contour levels (α-shells in the 3-D case) or density heights (in the 1-D and 2-D cases). This simple device makes it very easy to spot the number and location of high-density regions by focusing on one color.

The second use of color is for identification when grouping. This is most similar to the ordinary use

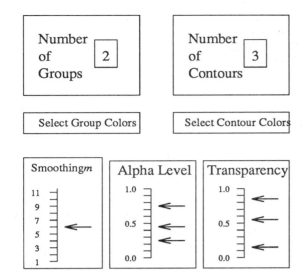

Figure 4: Sample Control Panel For the Density Grand Tour.

of color when brushing scatter plots. Here, each group's density estimate is colored uniquely for identification. The question of the best (or most effective) use of color when multiple α-shells and groups are desired is not fully answered. Clearly, things can get very busy on the screen if there are lots of groups and several α-shells displayed.

A full control panel can be imagined. There would be sliders for the number of α-shells; a slider for the values of each α level; menus for the color of an α-shell and group; and finally, a slider for transparency of each α-shell. An example control panel is shown in Figure 4.

7. Discussion

As with all graphical devices, experimentation is the true test. One can contemplate even more esoteric objects such as a 4-D DGT, which would display an 3×3 or 4×4 array of 3-D DGT slices going along. This idea can be implemented as with the 3-D DGT by having have 9 or 16 times as many single processors, each running an ASH and providing a single view to an output device.

Objects that allow brushing of bivariate histograms might be a natural extension of brushing scatter diagrams. Scott (1988) examined the theoretical benefits of using hexagonal bins, while Carr (1990) found hexagonal bins graphically appealing.

A more sophisticated idea would be to replace the brushing tool (typically a rectangle) by a "kernel," whose location is controlled by the mouse, and whose shape can be set by mouse interactions beforehand (as is the size and shape of the rectangular brush

in XGobi currently). This would provide "density brushing" as opposed to simple grouping. A conditional density might be the output in a coupled window.

With many dimensions, the grand tour is exhausting before all interesting directions can be exhausted. Projection pursuit is attractive for low dimensions, but new technology such as Terrell's informative components analysis (Terrell, 1985) provides endless possibilities for subspaces to tour, as well as ending subspaces. We plan to examine such ideas and look forward to ever more powerful graphical exploration tools capable at the high end.

Acknowledgments

Paper presented at the Annual Meetings of the ASA, Orlando, Florida, August 15, 1995. The author would like to thank John D. Salch for assisting in the creation of the video tape and figures in this paper, and Keith Baggerly for his comments.

David W. Scott is Professor, Department of Statistics, Rice University, Houston, TX 77251. This research was supported in part by the National Science Foundation under grant DMS-9306658 and the National Security Agency under grant MOD 9086-93.

References

Asimov, D. (1985), "The Grand Tour: A Tool For Viewing Multidimensional Data," *SIAM J. Scient. Stat. Comp.*, **6**, 128-143.

Buja, A. and Asimov, D. (1986), "Grand Tour Methods: An Outline," *Comp. Sci. Stat: Proc 17th Symp. Interface*, pp. 63-67.

Carr, D.B. (1990), "Looking at Large Data Sets Using Binned Data Plots," PNL-7301, Battelle Labs, Richland, WA.

Cleveland, W.S. (1993), *Visualizing Data*, Hobart Press.

Cook, D., Buja, A. and Cabrera, J. (1993), "Projection Pursuit Indexes Based on Orthonormal Function Expansions," *J. Comp. Graphical Statist.*, **2**, 225-250.

Friedman, J.H. and Tukey, J.W. (1974), "A Projection Pursuit Algorithm for Exploratory Data Analysis," *IEEE Trans. Comp.* C-**23**, 881-890.

Hurley, C. and Buja, A. (1990), "Analyzing High-dimensional Data With Motion Graphics," *SIAM J. Scient. Statist. Comp.*, **11**, 1193-1211.

Jee, J.R. (1987), "Exploratory Projection Pursuit Using Nonparametric Density Estimation," *ASA Proc. Statist. Comp. Sect.*, pp. 335-339.

Scott, D.W. (1985), "Averaged Shifted Histograms: Effective Nonparametric Density Estimators in Several Dimensions," *Ann. Statist.* **13**, 1024-1040.

Scott, D.W. (1988), "A Note on Choice of Bivariate Histogram Bin Shape," *J. Official Statistics* **4**, 47-51.

Scott, D.W. (1991), "On Estimation and Visualization of Higher Dimensional Surfaces," In *IMA Computing and Graphics in Statistics*, P. Tukey and A. Buja, Eds., pp. 187-205, Springer-Verlag, New York.

Scott, D.W. (1992), *Multivariate Density Estimation: Theory, Practice, and Visualization*, New York, John Wiley.

Scott, D.W. and Salch, J.D. (1995), "Demonstration of the Density Grand Tour With Brushing Extensions," Video Tape, Department of Statistics, Rice University, Houston.

Scott, D.W. and Thompson, J.R. (1983), "Probability Density Estimation in Higher Dimensions," *Proc. 15th Interface*, J.E. Gentle, ed, pp. 173-179, North-Holland.

Stuetzle, W. (1987), "Plot Windows," *J. Amer. Statist. Assoc.*, **82**, 466-475.

Swayne, D.F., Cook, D., and Buja, A. (1991), "XGobi: Interactive Dynamic Graphics in the X Window System with a Link to S," In *ASA Proc. Sect. Statist. Graphics*, pp. 1-8.

Terrell, G.R. (1985), "Projection Pursuit Via Multivariate Histograms," TR 85-7, Dept. Math Sciences, Rice University, Houston.

Wegman, E.J. (1990), "Hyperdimensional Data Analysis Using Parallel Coordinates," *J. Amer. Statist. Assoc.*, **85**, 664-675.

Wegman, E.J., Carr, D.B., Luo, Q. (1993), "Visualizing Multivariate Data," In *Multivariate Analysis: Future Directions*, pp. 423-466.

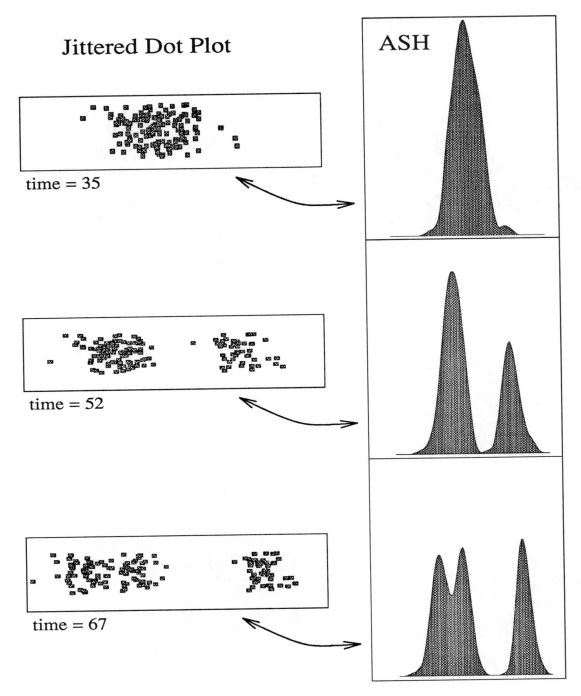

Figure 1: Three Frames From the Density Grand Tour of the Iris Data.

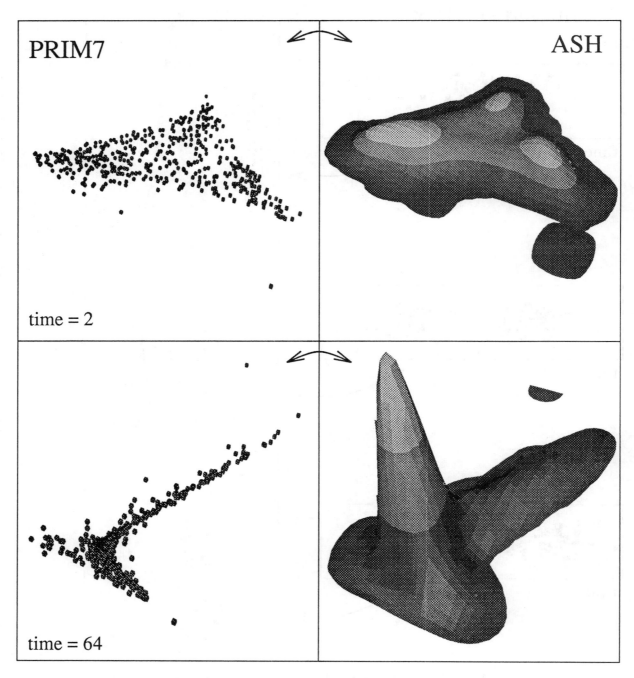

Figure 2: Two Frames From the Density Grand Tour of the PRIM7 Data.

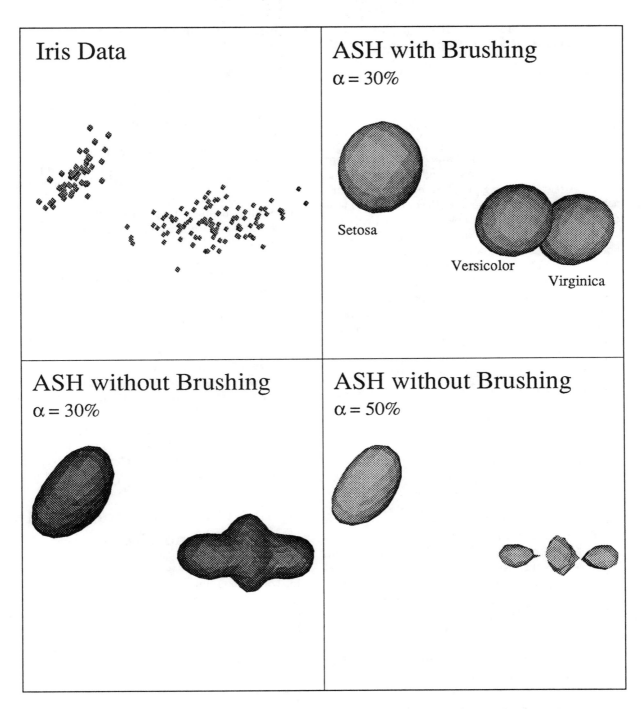

Figure 3: Frame From DGT of the Iris Data With and Without Brushing.

AN INTRODUCTION TO THE VISUALIZATION OF HIGH-DIMENSIONAL FUNCTIONS THROUGH FUNCTIONAL ANOVA

Charles B. Roosen and Jerome H. Friedman, Stanford University
Charles B. Roosen, Department of Statistics, Stanford University, Stanford, CA 94305

Keywords: Integration, Multivariate Function Estimation, MARS, CART

Abstract

In recent years the statistical and connectionist communities have developed many high-dimensional methods for regression (e.g. MARS, feedforward neural networks, projection pursuit). Users of these methods often wish to explore how particular predictors affect the response. One way to do so is by decomposing the model into low order components through a functional ANOVA decomposition and then visualizing the components. Such a decomposition, with the corresponding variance decomposition, also provides information on the importance of each predictor to the model, the importance of interactions, and the degree to which the model may be represented by first and second order components.

This paper describes such an ANOVA decomposition, presents methods for calculating the components, and gives an example of such a decomposition when modelling a real dataset.

1 Introduction

With the increase in computing power over the past decade, both the statistical and connectionist communities have developed a wide range of high-dimensional regression techniques. In statistics, this includes techniques such as CART (Breiman et al., 1984), MARS (Friedman, 1991), and projection pursuit (Friedman and Stuetzle, 1981). The connectionist community has developed a vast array of neural network methods (Hertz et al., 1991). One criticism of these methods is that users often view them as black boxes. The user may be interested not just in accurate prediction, but in interpreting the effect of the various variables on the response. A next step in the development of these methods is the improvement of diagnostic, inferential, and interpretation techniques for these models.

We focus here on one set of interpretation techniques based on examining low-dimensional components of a high-dimensional function. We may think of a fitted prediction model as a function $f(\mathbf{X})$. Alternately, we may be interested in some closed form function $f(\mathbf{X})$, such as a complicated likelihood function. The approach is to decompose our function $f(\mathbf{X})$ into first order effects depending on a single variable, second order cross terms depending on pairs of variables, and higher order terms depending on three or more variables. This is done by way of an Analysis of Variance (ANOVA) decomposition analogous to the discrete ANOVA familiar to statisticians. Although theory can be developed to produce components of order greater than two, we focus on one and two dimensional pieces so that they may be easily plotted and interpreted.

Gu and Wahba (1992) develop an ANOVA decomposition for smoothing splines in which the components are projections onto orthogonal subspaces whose elements satisfy certain side conditions. Stone (1994) describes a decomposition similar to the one given here, in which the components are orthogonal with respect to expectation. He focuses on tensor products of polynomial splines, and provides numerous convergence results for this class of functions.

First we describe the ANOVA decomposition and its properties. Next we discuss techniques for constructing the effects and variance components, using either analytic or numerical integration depending on the form of the model. Then we provide some example effect plots.

2 ANOVA Decomposition Definition and Properties

Owen (1992) describes an ANOVA decomposition for continuous functions which has nice orthogonality and variance decomposition properties. He refers to Efron and Stein (1981) and others cited by them before presenting the following definitions and properties. These authors rely upon a product measure for the predictor density, which we in turn use here.

2.1 Definition

We define our effects as conditional expectations adjusted to have mean zero, with low order effects removed from higher order effects.

Let $\int f^2(\mathbf{X})dF < \infty$, where $dF = \prod_{j=1}^{d} p_j(X_j)dX_j$ is a product density. Let $u, v \subseteq D = \{1, ..., d\}$ denote subsets of the axes of the predictor space. We use dF_u for integration with respect to the axes in u, leaving a function defined over the axes in $D - u$. That is, $dF_u = \prod_{j \in u} p_j(X_j) dX_j$. Similarly, dF_{-u} indicates integration with respect to the complement axes $D - u$.

The general form of an effect is

$$\alpha_u = \int \left\{ f - \sum_{v \subset u} \alpha_v \right\} dF_{D-u}.$$

Thus the grand mean is

$$\mu = \int f(\mathbf{X}) dF.$$

The main effects are

$$\alpha_j(X_j) = \int (f - \mu) dF_{-j}, \quad 1 \le j \le d.$$

The cross effects (second order effects) are

$$\alpha_{jk}(X_j, X_k) = \int (f - \mu - \alpha_j - \alpha_k) dF_{-jk}, \quad j \ne k.$$

Note that our main effects are conditional expectations adjusted to have mean zero

$$\alpha_j(x_j) = E\{f(\mathbf{X})|X_j = x_j\} - \mu,$$

while higher order effects are linear combinations of conditional expectations.

2.2 Properties

2.2.1 Additivity

With these definitions, we may expand the function f as a sum of effects:

$$\begin{aligned} f(\mathbf{X}) = \quad & \mu \quad + \quad \sum_j \alpha_j(X_j) + \sum_{j < k} \alpha_{jk}(X_j, X_k) \\ & + \quad \cdots + \alpha_{1\cdots d}(X_1, ..., X_d). \end{aligned}$$

2.2.2 Orthogonality

This continuous ANOVA has many properties analogous to those of the familiar discrete ANOVA. It is known that

$$\int \alpha_u dF_j = 0, \quad \text{if } j \in u$$

from which it follows that

$$\int \alpha_u \alpha_v dF = 0, \quad u \ne v$$

by writing $dF = dF_j dF_{-j}$ for $j \in u \cup v - u \cap v$. That is, the effects are orthogonal with respect to expectation.

2.2.3 Variance Components

Using the orthogonality of the components, it can be shown that

$$\int f^2 dF = \sum_{u \subseteq D} \int \alpha_u^2 dF.$$

Given this sum of squares decomposition, we can decompose the variance of f into variance components for each of the effects. By simply subtracting μ^2 from both sides of this equation and noting that $E(\alpha_u) = 0$ implies $\text{Var}(\alpha_u) = E(\alpha_u^2)$, we get

$$\text{Var}(f) = \sum_{u \ne \emptyset} \text{Var}(\alpha_u).$$

3 Predictor Density

We currently use either a uniform density over a hyperrectangle, or the product of the empirical marginals for observed training data (when we are visualizing a fitted model) as the predictor density. A product measure is highly desirable, as it yields the additivity and orthogonality properties mentioned above plus allows computational simplifications. The density used will affect the interpretation of the resulting effects. The ramifications of using these product densities and alternative non-product densities with analogously defined effects is discussed in a companion manuscript (Roosen, 1995).

4 Effect Estimation

To estimate main and cross effects, we must integrate the function numerous times, each time integrating out all but one or two variables. For some types of models it is feasible to perform the integration analytically. In other cases, numerical integration must be used.

4.1 Analytic Integration

Models built from axis-oriented components are particularly desirable when constructing the decomposition, as they may be analytically integrated over the predictor space. Such models include MARS, CART, the Π-method (Breiman, 1991), and tensor product splines (Gu and Wahba, 1991).

Models with a basis function representation make the integration particularly tractable. For concreteness, we will discuss the integration of a MARS model. (Note that the main MARS paper (Friedman, 1991) describes an ANOVA decomposition similar in spirit to that presented here, but lacking the nice additivity and orthogonality properties present in the decomposition under discussion.)

A MARS model is a model of the form

$$f(\mathbf{X}) = \beta_0 + \sum_{i=1}^{M} \beta_i \prod_{j=1}^{m_i} [s_{ji} \cdot \{x_{\nu(j,i)} - t_{ji}\}]_+$$

where $s_{ji} = \pm 1$, $x_{\nu(j,i)}$ is a particular predictor depending on i and j, and each predictor appears at most once in each product. By linearity of expectation, we need only consider integrating each tensor product. Due to the use of a product measure, we may group each basis with the corresponding marginal density when integrating, and we need only find a series of one-dimensional integrals. Hence the tensor product form of the model coupled with the product form of the density allows us to avoid the curse of dimensionality by reducing the multivariate integration problem to a set of univariate integrals.

Integration against a uniform marginal is standard integration of the basis function over a fixed range. When integrating against an empirical marginal, we represent the marginal density as a sum of point masses at the observed marginals. The integral of the basis function is then the sample mean of the basis function evaluated at these observed values.

4.2 Numerical Integration

Models which lack the axis-oriented structure are much more difficult to integrate analytically. These models include ridge-oriented functions (e.g. projection pursuit and many feedforward neural nets), and radially-oriented functions (e.g. radial basis functions). For these functions we may use numerical integration to derive the desired integrals.

The basic approach we consider is to evaluate the function over some design, project the data down onto the desired margins, and estimate the conditional expectation at a marginal design point by the sample mean at that point. To integrate against a uniform, we take the L levels on each axis to be L equispaced values. To integrate against empirical marginals, we instead take L equispaced quantiles. Note that if our marginals are uniform these procedures are equivalent.

Using a full balanced design is often computationally infeasible when d is large. Owen (1992) recommends the use of randomized orthogonal arrays in this context as designs which achieve a nice tradeoff between sample size and accuracy. We use randomized orthogonal arrays of strength three. (In these designs L^3 points are placed in the hypercube such that in each bivariate projection L points project down onto each of L^2 grid points, and in each univariate projection L^2 points project down onto each of L grid points. The points are placed such that the L points projected onto a bivariate grid point appear as random uniforms with respect to the complement variables). The strength three arrays appear to be sufficiently strong to estimate the main and cross effects, but are not as satisfactory when estimating variances. Identifying suitable alternative designs is an important area of current research.

5   The Variance Decomposition

Another summary provided by the ANOVA decomposition is the variance decomposition analogous to the sums of squares decomposition in discrete ANOVA. We can decompose the variance into variance components for the main effects, cross effects, and higher order terms. Note that here variance refers to the variation of the fitted function rather than some sort of measure of noise. To obtain a variance component using analytic integration, we first form the desired effect and then obtain its variance integrating over the predictors which are its arguments. With numerical integration, we essentially take the sample variance of the estimated values within each effect.

By looking at the proportion of the total variance of the function explained by the main and cross effects, we can examine the degree to which the function may be represented by first and second order pieces. If the decomposition is produced analytically and the function is intrinsically first or second order, then the first and second order variance components will sum to the total variance. If the function is instead of higher order, these components will sum to some value less than the total variance.

If the decomposition is produced by numerical integration, more care must be taken to avoid misleading results due to the nature of the design. A design which is only strong enough to capture first and second order effects will be unable to detect higher order effects, and hence the variance decomposition will suggest that only first and second order effects exist. Also, note that in numerical integration over orthogonal arrays, signal on an interaction appears as noise on the complement variables. This apparent noise will appear as variance in the variance decomposition. Due to this, the variance components with numerical integration may actually sum to more than the true variance.

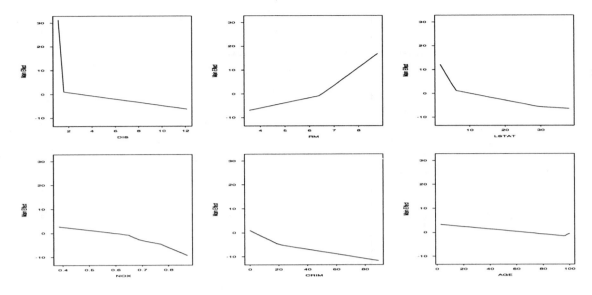

Figure 1: MARS Main Effects (DIS, RM, LSTAT, NOX, CRIM, AGE)

6 Example

We include a number of plots to display some of the characteristics of this methodology when applied to a real dataset. Space limitations prevent a complete analysis.

Data The Boston housing price data (Harrison and Rubinfeld, 1978) consists of 506 observations on 13 continuous predictors and one binary predictor, with the response representing median owner-occupied home values (MEDV) in Boston neighborhoods. This data is available from `Statlib`. The 9 predictors for which we display plots are:

CRIM per capita crime rate by town

NOX nitric oxides concentration (parts per 10 mil)

RM average number of rooms per dwelling

AGE proportion of units built prior to 1940

DIS weighted dist to 5 Boston employment centers

TAX full-value property-tax rate per $10,000

PTRATIO pupil-teacher ratio by town

B $1000 \cdot (Bk - 0.63)^2$ where Bk is the proportion of blacks residents by town

LSTAT % lower status of the population

Models Using S, we fit a MARS model and a tree model with MEDV as the response and all 13 predictors as continuous variables. The MARS model used a call to MARS 3.5 specifying at most second order interactions and 30 basis functions (i.e. mi=2 and nk=30), returning a model with 25 basis functions. The tree model was fit using `tree()` with the default parameters, giving a tree with 20 terminal nodes.

To form effects, the product of empirical marginals was used as the predictor density. With MARS analytic integration was performed, while with `tree()` we used numerical integration with 25 levels.

Main Effects The 6 main effects for each model with the highest contributions to the total model variance are displayed in Figures 1 and 2. In order of main effect variance, MARS uses DIS, RM, LSTAT, NOX, CRIM, and AGE, while `tree()` uses RM, LSTAT, DIS, NOX, PTRATIO, and TAX.

Note that MARS and `tree()` emphasize the same variables, and have simliar trends. Artifacts of the model classes are evident, with the MARS plots consisting of broken lines and the `tree()` plots consisting of step functions. Also note the slight jiggle in the `tree()` plots due to numerical integration.

Cross Effects We display the two MARS cross effects with the highest variances components (DIS and B, NOX and RM) in Figure 3, and the two such `tree()` cross effects (DIS and LSTAT, RM and LSTAT) in Figure 4. The MARS cross effect of DIS

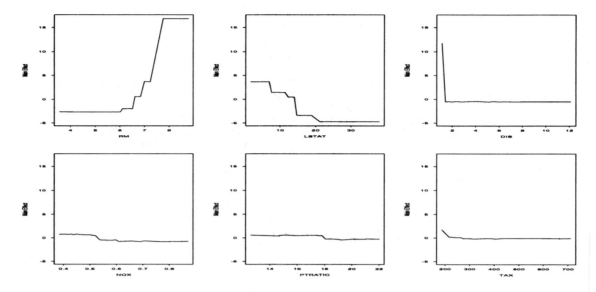

Figure 2: Tree Main Effects (RM, LSTAT, DIS, NOX, PTRATIO, TAX)

and B is hard to interpret without a better understanding of the definition of B. The NOX and RM cross effect is a classic interaction in which all houses are devalued by high NOX, but houses with many rooms particularly so. The tree() RM and LSTAT interaction is complex, and reflects the step form of a tree model. The tree() DIS and LSTAT interaction shows that in downtown the class status of a neighborhood has a big effect on housing price, and evinces some jiggle from numerical integration.

Variance Explained The eight MARS terms plotted here account for 86% of the total variance in the MARS model, with the remaining 7 main and 76 cross terms accounting for the other 14%. (Since mi=2 we know there are no higher order terms.) The eight tree() terms account for 89% of the total tree() model variance. Hence a large proportion of the functional variation of the models may be displayed using a small subset of the effects.

7 Extensions

Numerous extensions to the basic methodology described here have been developed. These provide a great deal of information complementary to that in the basic ANOVA decomposition. They are discussed in the companion manuscript (Roosen, 1995) and include:

- **Conditional variance plots** which provide information on regions of interaction.

- **F-test analogues** to assess variable importance.

- **Resampling methods** to assess the stability of effects.

8 Conclusions

This paper describes a functional ANOVA decomposition which is useful for examining the low-dimensional components of a high-dimensional function, and discusses general techniques for obtaining such a decomposition. Much additional work on this topic is described in a companion manuscript, along with directions for future research.

S Algorithms

Algorithms implementing these techniques in the S statistical programming language are available from the first author (charles@playfair.stanford.edu).

Acknowledgments

The authors thank Art Owen and numerous other colleagues who have provided valuable advice and support.

References

[1] Breiman, L. (1991), "The Π–Method for Estimating Multivariate Functions from Noisy Data," *Technometrics*, 33, 125–143.

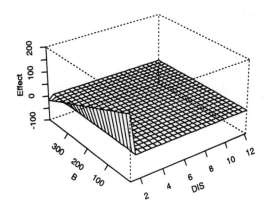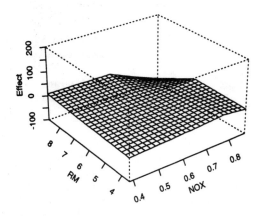

Figure 3: MARS Cross Effects (**DIS and B, NOX and RM**)

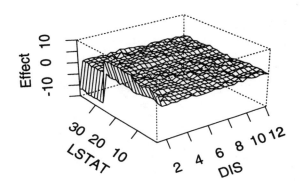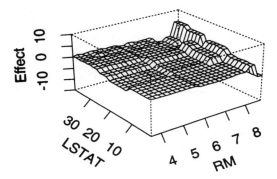

Figure 4: Tree Cross Effects (**DIS and LSTAT, RM and LSTAT**)

[2] Breiman, L., Friedman, J. H., Olshen, R. A. and Stone, C. J. (1984), *Classification and Regression Trees*, Belmont, CA: Wadsworth.

[3] Efron, B. and Stein, C. (1981), "The Jacknife Estimate of Variance," *Annals of Statistics*, 9, 586–596.

[4] Friedman, J. H. (1991), "Multivariate Adaptive Regression Splines," *The Annals of Statistics*, 19, 1–141.

[5] Friedman, J. H. and Stuetzle, W. (1981), "Projection Pursuit Regression," *J. Amer. Statist. Ass.*, 76, 817–823.

[6] Gu, C. and Wahba, G. (1991), Comments on "Multivariate Adaptive Regression Splines," *Annals of Statistics*, 19, 115-123.

[7] Gu, C. and Wahba, G. (1992), Smoothing Splines and Analysis of Variance in Function Spaces, Technical Report No. 898, Department of Statistics, University of Wisconsin.

[8] Harrison, D. and Rubinfeld, D. L. (1978), "Hedonic Prices and the Demand for Clean Air", J. Environ. Economics & Management, 5, 81–102.

[9] Hertz, J., Anders, K. and Palmer, R. G. (1991), *Introduction to the Theory of Neural Computation*, New York: Addison-Wesley.

[10] Owen, A. B. (1992), "Orthogonal Arrays for Computer Experiments, Integration and Visualization," *Statistica Sinica*, 2, 439–452.

[11] Roosen, C. B. (1995), Visualization and Exploration of High-Dimensional Functions Using Functional ANOVA, Doctoral Thesis, Department of Statistics, Stanford University.

[12] Stone, C. J. (1994), "The Use of Polynomial Splines and Their Tensor Products in Multivariate Function Estimation", *Annals of Statistics*, 22, 118–184.

DYNAMIC TECHNIQUES FOR VISUALIZING TRANSPORTATION FLOW DATA

David T. Hunt, ALK Associates, Inc. and Alain L. Kornhauser, Princeton University
David T. Hunt, ALK Associates, Inc., 1000 Herrontown Road, Princeton, NJ 08540

KEY WORDS: Network Visualization, Interactive Graphics, Traffic Animation, Traffic Assignment

1. Introduction

Transportation involves the flow of people and goods. It is the ensemble of a large number of individual demands for, and supplies of, transportation over both space and time that give rise to its most challenging problems. While summary statistics give some insight into these problems, a deeper appreciation and potential solutions only arise from the analysis of detailed measures that are used to characterize the travelers and their carriers. These analyses generate a deluge of data gained from either observations or simulations. In their raw form, these data can easily be overwhelming. To improve information absorption, visualization has been a standard tool used by transportation professionals to encapsulate this vast amount of data and transform it into operational ideas, concepts and solutions. Cartographic maps delineating transportation routes and demand locations were the earliest forms of visualizations. Since at least the mid nineteenth century traffic density maps have been used to depict directional flows of people and goods. More recently, the computer has progressively delivered more ability to capture operational data, generate new data through simulation, and host sophisticated visualization tools that readily transform those data into information. This paper presents new dynamic and real-time visualization techniques to vividly display the spatial and dynamic character of transportation.

2. Dynamic Flow Maps

This section explores one of the most widely used transportation visualization techniques: traffic density maps. Section 2.1 describes a traffic density map. Section 2.2 describes path generation and traffic assignment algorithms and then introduces a new algorithm which makes it possible to quickly move from the link-based flow map back to the origin-destination-based data. Section 2.3 uses this algorithm to dynamically brush and link flow maps.

2.1 Introduction to Flow Maps

The traffic density map, or flow map, displays both volume and spatial patterns. A coordinate system (preferably geographically correct), provides the location for the start and end of a trip. A band is constructed with a width proportional to the amount of traffic moving between these two locations. If the bands are constructed along a network (such as a road or rail system), the flow map can effectively represent the movement of goods from multiple origins to multiple destinations. As a general convention, the right-hand rule is applied to two directional flows. Imagine standing on the center line of a road. Traffic moving from back to front would be displayed on the right-hand side of the center line, while traffic moving from front to back would be displayed on the left-hand side. Thus, for an east-west road, eastbound traffic is displayed on the bottom.

Figure 2.1 contains an example of a flow map using the 1993 Interstate Commerce Commission Railroad Waybill Sample, Public Use File. The waybill sample is collected annually by the ICC from the terminating railroad. It is approximately a 2 1/2 percent sample factored to represent total railroad movements. The public use file protects sensitive railroad data by aggregating locations to Business Economic Areas (BEA) and excluding geographic information if it is possible to identify an individual railroad. Figure 2.1 displays the net tonnage that moved over the railroads, but due to ICC restrictions the volumes are limited to approximately 50% of the total 1993 U.S. rail tonnage.

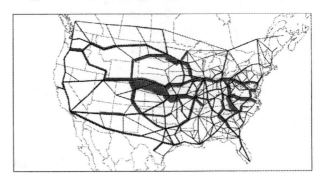

Figure 2.1 -- 1993 ICC Rail Waybill Sample, Public Use File. The maximum link volume is 83 million net tons.

Flow maps are certainly not a modern invention. Charles Joseph Minard effectively used flow maps in the mid-1800s to display the movement of both people and goods. In 1861 Minard displayed the movement of troops from the Polish-Russian border to Moscow and back to the border during Napoleon's Russian campaign of 1812. [Tufte, page 41]. Minard later produced a flow map showing the worldwide export of French wines in 1864. [Tufte, page 25] As with most statistical graphics, computers have allowed flow maps to be constructed quickly and efficiently. But these computer representations are still static and differ little from the flow maps produced by Minard over one hundred years ago.

2.2 Matching O-D and Volume Matrices

Probably the most fundamental computational modeling activity done in transportation is to reconstruct the route, the sequence of n links, taken by a trip from an origin to a destination. Because of the data intensity required to depict even one route from each origin to each destination, such data are rarely collected from actual observations. Furthermore, any change in the transportation supply, which is reflected by changes in link characteristics, may generate a new route, as would congestion by changing speed and variations of driver/shipper preferences. Thus, an efficient means of generating routes is necessary to transform O-D demand into transportation facility utilization. Utility maximization theory is used as the underlying principle to substantiate the algorithmic route generation process. In its simplest form, it assumes that all travelers/shippers have the same utility function and that the link cost function is a constant. This formulation leads to the well-known least-cost path linear-network problem of finding the sequence of links that minimizes the sum of link costs, L, while satisfying the boundary conditions that the path start at node O and end at node D. More detailed formulations consider the equilibrium problem of either minimizing costs for individual trip makers (user optima) or all trip makers (system optima) where the cost for each individual trip maker is a function of volume. This leads to a non-linear network equilibrium problem that is much more intensive computationally. While it is indeed more desirable to solve the non-linear problem, the individual non-linear congestion sensitive utility functions are not well known. Often, the additional computational burden can not be substantiated because of the error contained in the non-linear utility function.

A flow map is most often generated using an all-or-nothing assignment of the traffic originated and terminated at every node to the links forming the least-cost path between the origin and destination nodes. Starting with an origin-destination matrix (containing origin node (O), destination node (D), and volume of traffic from O to D), the traffic assignment algorithm creates one least-cost tree from every origin. These trees are used to construct a volume matrix containing link number (with endpoints A and B), the total volume of traffic moving in the AB direction, and the total volume of traffic moving in the BA direction. The traffic moving over any individual link consists of an aggregation of traffic from several origin and destination pairs. A problem has been trying to quickly disaggregate the traffic volume data back to the constituent O-D data. For example, Figure 2.1 contains 170 origins, 174 destinations, 338 links, and 5361 unique origin-destination pairs. Thus, the O-D matrix has 5361 rows and three columns (origin, destination, volume). The volume matrix has 338 rows and three columns (link number, volume from A to B, and volume from B to A). If the volume matrix were expanded to include origin-destination information, then every link would have to be listed separately with the O-Ds that traversed that link. If the average number of links in a path is 20, then the volume matrix would increase from 338 rows and 3 columns to 107,220 (5361x20) rows and 5 columns (link number, origin, destination, AB volume, BA volume). Not only does storing these data for each alternative in an application become a challenge, but one can only absorb information from the data on a query basis. Thus, instead of storing a lot of data that will probably never be analyzed, what is needed is a fast

algorithm that reconstructs this information for only the link(s) of interest.

The following figure contains a simple network for illustrating an algorithm to identify the origin destination pairs moving over a specific link.

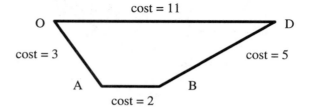

Identification of the traffic moving over link AB requires running two least-cost path algorithms. First run a least-cost path from node A to nodes O and D. Retain the costs. Next, run a second least-cost path from B to O and D. If the least-cost path cost to move from O to D is equal to the lower value of the cost from A to O and B to O plus the lower value of the cost from A to D and B to D plus the cost of link AB, then the traffic from that origin and destination must travel over link AB, (or in the case of a degenerate network, a path of equal cost). In the above example, the least-cost to move from O to D is 10. This is equal to:

$$\text{Min} (C_{OA}, C_{OB}) + \text{Min} (C_{AD}, C_{BD}) + C_{AB} =$$

$$\text{Min} (3,5) + \text{Min} (7,5) + 2 = 10$$

Thus, traffic moving from O to D would traverse link AB. This algorithm is described in more detail in Appendix A.

2.3 A dynamic flow map

Since the algorithm described in Section 2.2 involves using the cursor to dynamically alter the graphic by highlighting specific elements of the map, the term "brushing" is borrowed to refer to this procedure. Brushing involves real-time alteration of selected data objects on a computer graphics display through direct manipulations with the cursor. This process is often applied to multivariate data sets represented by scatterplots and scatterplot matrices. [Becker and Cleveland, 1987] Brushing provides a means for separating and identifying data objects without regenerating the entire graphic or losing the information supplied by the non-highlighted data.

Figure 2.2 displays the same 1993 railroad data shown in Figure 2.1. An arrow now points to a link in southeastern Wyoming. This link transports more tonnage than any other link in the network. Most of this tonnage is coal moving out of the Powder River Basin. This link also transports import and export intermodal traffic between the west coast ports and eastern locations, plus an assortment of general merchandise traffic. Looking at Figure 2.2, it is impossible to determine the origins and destinations for the data moving over this link. Is the traffic on this link local to a few selected origins and destinations or does it fan out over several markets? Do the large coal flows out of Powder River move to east coast locations?

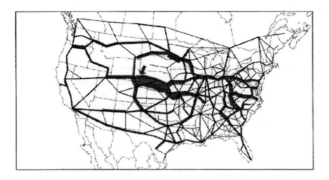

Figure 2.2 -- Arrow highlights the link to be brushed.

Figure 2.3 shows the results of brushing the link highlighted by the arrow in Figure 2.2. The tree structure represents the full routing for all of the origin-destination pairs that utilize the selected link in their path. It is now obvious that this link transports traffic for several origin-destination pairs and serves a surprisingly large number of eastern locations. It is also obvious that while a large number of origin-destination pairs are served, the tonnages are relatively small indicating that little Powder River coal moves by rail into the eastern US. Some of the larger flows, especially in southern Illinois, appear to end at river locations and are most likely rail-barge transloading facilities for coal movements down the Mississippi. (The scales on Figures 2.1, 2.2, and 2.3 have been held constant to permit direct comparison of the flows.)

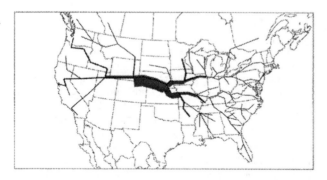

Figure 2.3 -- All traffic traveling over the brushed link.

Brushing has greatly enhanced the amount of useful information that can be extracted from the flow data shown in Figure 2.2. Further interrogation and exploration of the flow data can be achieved by incorporating additional dynamic ability through the use of logical operators. These operators should provide basic selection, addition, and deletion functions along with "not", "blank", "and", and "or" capabilities. Consider an O-D matrix containing: origin, destination, attributes, volume, least-cost path cost. The attributes might include commodity, car type, or customer. Brushing a link will quickly point to the rows in this O-D matrix that traverse the link. Providing logical operators for the columns in the O-D matrix greatly enhance the ability to identify specific movements. For example, display all of the non-coal traffic originating or terminating in Los Angeles and traversing the link highlighted in Figure 2.2. These operators should not just apply to the O-D matrix, but should allow adding and deleting additional links. Selecting all traffic that travels over the link highlighted in Figure 2.2, but not over the link into Kansas City, would eliminate all of the flows to and from the southeastern US in Figure 2.3.

Linking the flow map to other graphics can further enhance the ability to discover useful phenomena in the data. For example, link a flow map displaying tonnages to a histogram containing revenue per ton-mile. Clicking on a link in the flow map will highlight both the flows on the flow map and the corresponding revenue per ton-mile in the histogram. Clicking on the largest revenue per ton-mile in the histogram would similarly highlight those movements in the flow map. [see Haslett, et. al. for an example of linking geographical maps and histograms]

3. Visualizing Traffic Flows Through Animation

Many of the most challenging transportation problems arise because transportation demand and supply are not uniformly distributed over time. Transportation flows exhibit surges in demand and disruptions in service that are both spatially and temporally distributed. Common examples are traffic accidents during the morning rush "hour". Demand is time-varying during that period and the road system's ability to serve that demand is curtailed by lane closings and "rubber necking". Simulation models that investigate alternative traffic management techniques are but one application in which vast amounts of time and spatially varying data need to be visualized. A straight forward approach to this visualization is gained by recognizing that trips over space and time are nothing more than trip filaments (trajectories) in a 3-dimensional space of geography (e.g., x-y, latitude-longitude, etc.) and time as depicted in Figure 3.1. Projections of all filaments onto the geography plane yield traffic density maps as discussed above. If instead, one borrows from 3-d computer graphic techniques a viewing frustum that "looks" down the time axis, clips the geographic plane to the desired viewing window and clips the time axis using a front clipping plane located at time t and a back clipping plane located at time $t+dt$, then what is projected on the viewing plane are the segment of those filaments that are in the geographic viewing window during the time interval $[t, t+dt]$. Further more, changing t by an amount Dt between each screen redraw, allows one to depict the movement of the trip over time and space. This construct has been extremely useful when used in conjunction with Silicon Graphics workstations that have specialized hardware for efficiently performing the frustum clipping and projection computations. Animations that simply manipulate the viewing frustum over time provide a temporal and spatial visualization of observational or simulated traffic data.

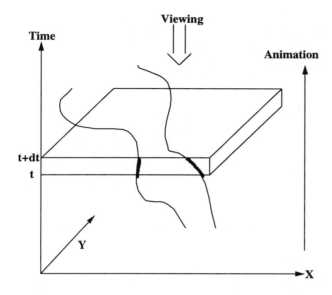

Figure 3.1 -- Two trips passing through a clipping plane.
The trip filaments are highlighted between time t and t+dt.

This approach was first used by Goodloe White and Alain Kornhauser to view airline activity on a typical day. The Flight Analysis Data System (FADS) takes as input airline flight schedules as available from sources such as the Official Airline Guide. These sources provide origin airport, origin time, destination airport, destination time, as well as other attributes such as carrier, flight number, and aircraft type. The airport and time data provide end-points for flight filaments that are "drawn" in 3-D space. When clipped and projected they depict the enroute location of aircraft at the appointed time. Progressive displays animate the movement of aircraft, as shown in Figure 3.2. FADS provides an interface that allows the user to select data subsets and animation controls. [Kaiser and Kaplan]

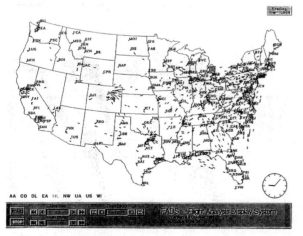

Figure 3.2: Flight Analysis Data System sample screen

The FADS system has been extended to depict highway traffic movement. Extensions include provisions for filaments that are composed of a sequence of links, filaments to be offset from a link centerline depending on direction of travel, and depiction of travel in individual lanes of a multi-lane highway. This system has been applied to a network consisting of New Jersey's major roadways to depict the movement of 1.8 million individual automobile trips during the morning commute period from 5:00 am to 11:00 am. In the 15 minutes that it takes to animate the 6 hours of traffic one begins to appreciate the intricacies of the spatial and temporal distribution of New Jersey's morning commute problems. In order to increase the value of the visualization, the user has control over the viewing window and can interactively select the display of summary link or node based statistics. One can also tag individual or groups of trips with a different color so that their progress through the network can be differentiated from other vehicles.

4. Essentially-Least-Cost Paths

In the previous sections of this paper, traffic was assigned to the path between the origin and destination that had the least-cost. Section 4.1 questions the accuracy of this and generalizes the concept of a least-cost path. Section 4.2 presents applications of essentially-least-cost paths.

4.1 Generating essentially-least-cost paths

Given an origin location and a destination location, traffic is generally assigned to the path with the least-cost between these two locations. The cost can be distance, time, monetary, distance weighted by road type, or any other metric that can be assigned to a link. The actual assignment is accomplished using a standard least-cost path algorithm, which is described in most standard operations research or network flow textbooks. [e.g.: Jensen and Barnes]

Consider the measurement accuracy associated with the costs used to route traffic over a least-cost path. Two routes between a large residential development and a shopping center might have a travel time cost of 17.1 and 17.3 minutes. The route with a cost of 17.1 would receive all of the traffic moving between these locations, while the route with cost of 17.3 would receive no traffic. Are the cost measurements accurate enough to perform an all-or-nothing traffic assignment based on this twelve second differential in route costs? These two paths are essentially-least-cost paths.

Appendix B describes an algorithm (based on the brushing algorithm) that will produce the subnetwork of all essentially-least-cost paths between an origin and destination that fall within a specified cost threshold. To find these essentially-least-cost paths, the least-cost path between the origin and destination is generated. A cost threshold (τ) is specified. Each link in the network is brushed to determine whether or not it can serve that origin-destination pair at a cost equal to or less than this increased "acceptable cost". If the brushed link can serve the origin-destination within the acceptable cost, the link is retained as part of the essentially-least-cost subnetwork. After each link has been brushed, the links in the subnetwork can be assembled into a set of unique paths that represent all combinations of paths between the origin and destination that are within the acceptable cost.

Figure 4.1 is based on the network used in Figures 2.1 through 2.3. It contains 192 nodes representing Business Economic Areas (BEAs) and 364 links connecting the BEAs. The dark lines display the subnetwork containing nineteen unique essentially-least-cost paths identified as being within two percent of the least-cost path between Savannah GA and Sacramento CA.

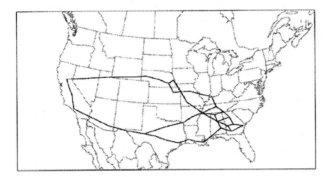

Figure 4.1 -- Essentially-least cost subnetwork within 2% of least cost path between Savannah and Sacramento. Contains 52 links and 19 paths (excluding paths that traverse a node more than once).

Figure 4.2 displays the same network, but now the dark lines represent the subnetwork containing 680 unique essentially-least-cost paths between Savannah and Sacramento that are within ten percent of the least-cost path.

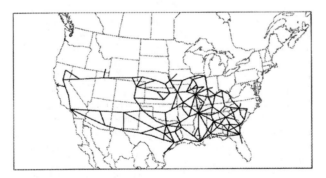

Figure 4.2 -- Essentially-least-cost subnetwork within 10% of least cost path between Savannah and Sacramento. Contains 168 links and 680 paths (excluding paths that traverse a node more than once).

Note the existence of "dead-end" links in Figure 4.2 (e.g.: the link into Idaho). These are part of a valid path in which the cost to traverse this link twice is still within the specified threshold. This points out an inconsistency: two travelers with the same destination and currently at the same location may have different route choices depending on their origin. Long distance paths may include "dead-end" links and other detours that are locally unacceptable. Alternatively, arriving at an intermediate location can open up new path opportunities not available from the origin. This inconsistency can be corrected by identifying all cycles in the network (within a threshold τ) and building the combinations of essentially-least-cost paths from these cycles. This procedure is more fully described in Hunt and Kornhauser.

Generation of the subnetwork of essentially-least-cost paths differs from the standard "k" shortest path algorithms which identify a set number (k) of shortest paths between an origin and destination. The "k" shortest path algorithms either alter the link cost data and generate a new least-cost path or systematically remove links from the network and rerun the least-cost path algorithm. [see Horowitz and Sahni] "K" shortest path algorithms are not designed to return all paths within a specified threshold.

4.2 Applications of essentially-least-cost paths

Traffic assignment was previously described as the assignment of all traffic between an origin and destination to the least-cost path in a network. In practice this method is easy to implement and provides a good aggregate representation of traffic movements when several origins and destinations are involved. It does not work well if a single, or few, origins and destinations are being studied. Several route choice models have been developed to predict the behavior of individuals traveling between locations. Most models assume that an individual will minimize their cost to travel through the network. The goal is to achieve a system equilibrium, where no individual trip maker can reduce their cost by switching to a different route. This problem is made more difficult by using non-linear costs that are a function of the volume of traffic on the link. Stochastic models that introduce a random component to route choice have also been developed for both linear and non-linear cost functions. [for a general reference see Kanafani]

A straight forward method for performing traffic assignment over the essentially-least-cost subnetwork introduced in Section 4.1, is to assign half of the traffic to each path at each decision point. Since all paths in the subnetwork have been declared equal, the actual path taken by a traveler can best be determined by the flip of a fair coin. Consider the following subnetwork:

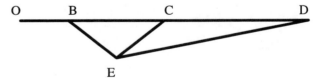

Assume that the paths in the subnetwork from origin O to destination D are {[O,B,C,D], [O,B,E,D], [O,B,E,C,D]}. The probability of using link OB is 1. At node B, assign half of the traffic to link BC and half to link BE. At node E, assign half of the traffic to link ED and half to link EC. Continuing with this assignment, the probability of using each link is: OB=1, BC=.5, BE=.5, EC=.25, ED=.25, CD=.75.

Within a transportation system (roadway, single railroad) the subnetwork of least-cost paths can be used to identify the most critical links in the network. Critical links are defined in this paper as those links common to all paths generated by the essentially-least-cost path algorithm. If a link appears in all of the essentially-least-cost paths for an O-D pair, that link is critical and the traffic between that O-D is captive to that link. If a link does not appear in all of the paths, then the traffic has an alternate available route within the cost threshold. It then becomes possible to identify the volume of traffic on a link that is captive to that link. Links with a low volume of captive traffic are potential candidates for rationalization.

Whereas alternate routings within an individual transportation system might be viewed as unnecessary, and in some instances wasteful, the lack of alternate routings between transportation systems can be viewed as monopolistic. Using the procedure described in the previous paragraph, it is possible to display competitive versus captive traffic in a region or corridor. For railroads, competitive traffic would have at least two alternative routes, within the cost threshold, on different railroads. This can be used to identify regions that might be affected by a railroad merger.

5. Summary

Visualization is the most efficient method for assimilating the enormous quantities of data generated by a transportation system. Techniques that permit dynamic interaction with the visual displays lead to an even deeper understanding of the patterns and relationships contained in the data. Brushing flow maps and computer animation of flows are two very effective methods for improved understanding of the origin-destination movement of goods and people.

This paper also questioned the practice of assigning traffic to one least-cost path when generating visual displays of transportation movement data. Since data are generally gathered on the trip origin and destination, but not the actual path, assignment of traffic to the incorrect path can lead to incorrect interpretations of the movement data. A new methodology was presented for generating all paths within a given threshold of the least-cost path. These were referred to as the essentially-least-cost paths. Applications include vehicle route choice and network analysis.

The authors would like to thank Stuart Smith of ALK Associates for his helpful comments. The authors would also like to thank Kenneth Rodemann of AT&T Bell Laboratories for identifying an omission in the essentially-least-cost path procedure.

References

Becker, Richard A., and William S. Cleveland, "Brushing Scatterplots", Technometrics, Volume 29, Number 2, May 1987.

Cleveland, William S., The Elements of Graphing Data, Revised Edition, Hobert Press, Summit New Jersey, 1994.

Cleveland, William S., and Marylyn E. McGill, Dynamic Graphics For Statistics, Wadsworth & Brooks/Cole, Belmont CA, 1988.

Haslett, John, Ronan Bradley, Peter Craig, Antony Unwin, and Graham Wills "Dynamic Graphics for Exploring Spatial Data With Applications to Locating Global and Local Anomalies", The American Statistician, Volume 45, 1991.

Horowitz, Ellis, and Sartaj Sahni, Fundamentals of Data Structures, Computer Science Press, Inc., Potomac MD, pages 324-327, 1976.

Hunt, David T., and Alain L. Kornhauser, "Assigning Traffic Over Essentially-Least-Cost Paths", to be presented at the 75th Annual Transportation Research Board, Washington, DC, January 1996.

Jensen, Paul A., J. Wesley Barnes, Network Flow Programming, John Wiley & Sons, Inc., New York, NY, 1980.

Kaiser, Steven and Eric Kaplan, The Flight Animation and Database System, Princeton University Class Project Report, 1990.

Kanafani, Adib, Transportation Demand Analysis, McGraw-Hill Book Company, New York, NY, 1983.

Tufte, Edward R., The Visual Display of Quantitative Information, Graphics Press, Cheshire, Connecticut, 1983.

Appendix A: Brushing Traffic Density Maps

Step 1: Given a set of origin nodes (O), destination nodes (D), and the volume between each O and D, perform a traffic assignment over the network. Append to each record the cost to move from O to D (C_{OD}).

Step 2: Draw the network on the computer screen. Use the cursor to brush a link in the network. Identify the endpoints (A and B) of the link and the cost to traverse the link (C_{AB}).

Step 3: Build a least-cost path tree from the A node on the brushed link to all other network locations. Store the resulting costs.

Step 4: Build a least-cost path tree from the B node on the brushed link to all other network locations. Store the resulting costs.

Step 5: If: $C_{OD} = \text{Min}(C_{OA}, C_{OB}) + \text{Min}(C_{AD}, C_{BD}) + C_{AB}$ then the origin-destination pair OD must traverse link AB, or in the case of a degenerate network, a link of equal cost.

Appendix B: Generating Essentially-Least-Cost Paths

Step 1: Select either one origin-destination (O-D) pair or a matrix of O-D pairs. This set could be all possible O-Ds, but this may require storage of a large matrix.

Step 2: Run one least-cost path from each origin to establish the least-cost path cost from the origin to the destination (C_{OD}). Inflate this cost by the desired amount (τ).

Step 3: "Brush" each link to determine the O-Ds defined in Step 1 that use that link. If the path cost is less than or equal to the inflated least-cost path cost, go to Step 4. If the path cost is greater than the least-cost path cost, brush the next link.

Rather than running 2 x L least-cost paths (where L is the number of links), run n least-cost paths (where n is the number of nodes) and temporarily store the least-cost path cost and predecessor links vectors. For example, brush link A-B and retain the cost and predecessor links vectors for A and B. Now brush link A-C. Since the cost and predecessor links vectors were stored for A, a least-cost path only needs to be run for C. The cost and predecessor links vectors for A can be erased after all links attached to A have been brushed. Note that it is not necessary to brush a link if at least one node on that link is beyond the cost threshold.

Step 4: Retain the "brushed" link as part of the essentially-least-cost subnetwork. After all links have been brushed, the links forming the subnetwork of essentially-least-cost paths will have been assembled.

Step 5: Using the subnetwork assembled in Step 4, construct all combinations of paths in this subnetwork. This can be accomplished with a branch and bound procedure where a branch is terminated if (a) the cost exceeds the cost threshold or (b) a node is repeated prior to reaching the destination. The surviving branches form all feasible paths between the origin and destination with a cost less than or equal to the cost threshold. This step will need to be repeated for each O-D pair identified in Step 1.

ESTIMATING DEPENDENCIES FROM SPATIAL AVERAGES

Audris Mockus, AT&T Bell Laboratories

Rm 1U-328, 1000 E. Warrenville Rd., Naperville, IL 60566. audris@research.att.com

Key Words: covariance function, aggregate data

Abstract

We consider modeling a space-time function using observations in the form of averages of the function over a set of irregularly shaped regions in space-time. Such observations are most common in applications where the data is gathered for administrative, political, geographic, or agricultural regions.

In order to estimate (predict) the value of this function we estimate the dependence structure of the underlying stochastic process. We propose a method to estimate the covariance function from the integrals of a stationary stochastic process. The method poses the problem as a set of integral equations which are then solved via least squares.

1 Outline

In many instances observations collected in an experiment or in a study represent averages of some quantity over a period in time and/or over a region in space. Examples of such observations/quantities include the number of cases of a disease, population size, income or other epidemiological, socio-economic, or physical quantity. The quantity of interest could be approximated by modeling it as a stochastic process of continuous space and time. In this paper we consider several ways to model a stochastic process using aggregated (over area and time) data. We considers ways to estimate the covariance function of such process and ways to predict the values of the processes.

The purposes of modeling are:

- To predict the function.

- To obtain statistical properties of the function.

- To display the function for visual inspection and exploratory analysis.

The last objective is more demanding than the first two as it requires efficient software implementation of results of the first three problems.

Eddy and Mockus (1995) describe Multivariate Interactive Animation System for Map Analysis (MIASMA) to demonstrate that the exploratory analysis of the considered type of data can not be successful without good modeling tools and the modeling is impossible without the exploratory tools. The software implementation of covariance function estimation described in this paper is being integrated into MIASMA system.

A problem of spatial estimation from aggregate data was considered by Tobler (1979) in the context of estimating population density. That paper contains a bibliography from the field of applied geography. The interpolation proposed by Tobler is a numeric solution of a Dirichlet's equation with boundary constraints. The method is designed to produce an interpolant under such geographic constraints as of lakes, mountains, and deserts. Dyn and Wahba (1982) describe a thin plate histospline approach for the similar problem. Available modeling methods include a kernel type smoothing method described in Eddy and Mockus (1994). None of those methods consider ways to estimate the dependence structure of the quantity that is being predicted.

The paper consists of three parts. First part describes datasets, second part will consider prediction, and the third part describes estimation of the dependence structure.

2 Datasets

The data was obtained from NNDSS (National Notifiable Disease Surveillance System). The dataset contains weekly by state reports on 57 diseases for the period between 1980 and 1994. There are 783 report weeks in this period and the reports are provided for 51 states, 3 territories and the New York City. We analyze 3 different popular (in the number of cases) diseases: Hepatitis A, Hepatitis B, and Tuberculosis. Short description of each disease is given below.

Hepatitis is a disorder involving inflammation of the liver. Hepatitis may be acute or chronic. Hepatitis A, once called infectious hepatitis, is the most common cause of acute hepatitis. Usually transmitted by food and water contaminated by human waste, infections can reach epidemic proportions in regions lacking adequate sanitary systems. Hepatitis B is spread mainly by blood or blood products. Type B virus is resistant to sterilization of instruments in hospitals. It often causes an initial episode of liver disease and occasionally leads to chronic hepatitis.

Tuberculosis, is an acute or chronic infectious disease caused by bacteria of the genus Mycobacterium, often called tubercle bacilli. Human infection with tuberculosis was one of the leading causes of death until antituberculous drugs were introduced in the 1940s. TB spreads when a person infected with the active form of the disease coughs infected droplets into the air, which are then inhaled by other people. In the mid-1980s there was a resurgence of tuberculosis in the United States. Since the mid-1980s, a drug-resistant strain has spread.

2.1 Initial inspection of the data

Table 1 lists 3 diseases with total reported counts over the period 1980-1994.

Disease Name	Total Counts
Hepatitis A	328964
Hepatitis B	280586
Tuberculosis	327441

Table 1: Total counts.

Time plots of average (over states) incidence rates (number of cases per week per 100,000 population) for 3 diseases in the United States are in Figure 1. All of the diseases are seasonal and have nontrivial trend over time. The behavior at the state level is even more complicated.

3 Prediction

In this Section we model the disease incidence rates as a smooth function over the territory of the United States and over considered time period. The information about function of interest is the number of cases of a disease for each state-month combination (it represents integrals of the incidence rate

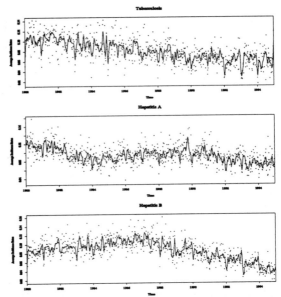

Figure 1: The time behavior of the 3 diseases in the US. Points show average weekly incidence rates. Lines are smoothed time series of points and are shown to emphasize trends. Tuberculosis is shown at the top, Hepatitis B - at the bottom

function with respect to population density over the territories of individual states and over calendar months).

More formally, we are interested in determining a function $f(x)$ from its integrals $z_i = \int_{A_i} f(x)dx$, where A_i's partition a finite set $A \subset R^d$. We model the function as a stochastic process. The simplest interpolant for the function f could be a piecewise constant function

$$\hat{f}(x) = \sum_i I_{x \in A_i} \frac{z_i}{\int_{A_i} 1dx} \qquad (1)$$

Since we intend to use the interpolant in animations the discontinuity of such function can be distracting and needs to be avoided (see Eddy and Mockus (1994)). Here we consider methods that produce a continuous interpolant.

Let $f(\mathbf{x})$ be a stochastic process with the index $\mathbf{x} \in A \subset R^d$. Let the observed values z of the single realization of the process be

$$z_i = \int I_{A_i}(x)f(t)dt, \qquad (2)$$

where I_{A_i}'s are indicator functions for disjoint sets $A_i \subset A$.

In case when the joint distribution of $(f(\mathbf{x}), \mathbf{z})$ is Gaussian the quantity $E(f(\mathbf{x})|\mathbf{z})$ is a well known linear function of $\mathbf{z}$ given by the formula for the conditional normal distribution. Let $\mu(\mathbf{x}) = E(f(\mathbf{x}))$, $\mu_i = E(z_i)$, $c_i(\mathbf{x}) = \text{Cov}(f(\mathbf{x}), z_i)$, $c_{ij} = \text{Cov}(z_i, z_j)$. Then

$$E(f(\mathbf{x})|\mathbf{z}) = \mu(\mathbf{x}) - (c_i(\mathbf{x}))(c_{ij})^{-1}((\mu_i) - \mathbf{z}). \quad (3)$$

In case when we just know the first two moments of the vector $(f(\mathbf{x}), \mathbf{z})$ Equation (3) defines best linear unbiased predictor for $f(\mathbf{x})$.

Equation (3) involves quantities μ and c that are unknown. The trend $\mu(\cdot)$ is usually assumed to be a constant and we could estimate $c_i(s), c_{ij}$ as described in the next section.

4 Dependence Structure

In this Section we describe methods to estimate geographic and time dependence structure for the stochastic process of disease incidence rates. There are a number of techniques to estimate mean and covariance functions from observations representing point values of a process. In our case the observations are in the form of integrals of the process f described in Section 3. In this section we assume that the process f is zero-mean and stationary. We are interested in the covariance function $\gamma(\mathbf{x}_1 - \mathbf{x}_2)$ of the process f. We obtain the estimate of γ by solving appropriate integral equations.

We describe an efficient algorithm to estimate γ for the case of an isotropic covariance function (isotropic means that $\gamma(\mathbf{x}_1 - \mathbf{x}_2)$ is a function of only $\mathbf{x}_1 - \mathbf{x}_2$), and the A_i's being polygons when $d = 2$ (d is the dimensionality of the argument space of the process $f(\mathbf{x})$) or prisms with a polygon base when $d = 3$. Methods to estimate the covariance function from point observations are different and can be found in, e.g., Cressie (1985), Marshall and Mardia (1985). Surprisingly, the estimation of the covariance function from integral observations has not been investigated.

In Subsection 4.1 integral equations are obtained for a covariance function. The method to estimate the covariance function from those equations is then presented. Two main obstacles to an efficient numeric solution are large number of equations and search in the space of positive definite functions (we want our solution to be a valid covariance function). We address the first problem by designing a fast algorithm to compute the integral of interest and the second problem by re-expressing the isotropic covariance function in terms of a nondecreasing function.

4.1 Estimation Equations

Let f be a stationary zero-mean Gaussian process on a set $A \subset R^d$ having unknown covariance function γ. Let $z_i = \int_{A_i} f(\mathbf{x}) d\mathbf{x}$, where $A_i \subset A$, $i = 1, \ldots, N$. Then the vector $(z_1, \ldots, z_N)$, where $z_i = \int_{A_i} f(\mathbf{x}) d\mathbf{x}$, $A_i \subset A$, has a multivariate Normal distribution with expected value 0 and covariance matrix

$$\text{Cov}(z_i, z_j) = \int_{A_i} \int_{A_j} \gamma(\mathbf{u}, \mathbf{v}) d\mathbf{u} d\mathbf{v}$$

The products $z_i z_j$ approximate the integrals $\int_{A_i} \int_{A_j} \gamma(\mathbf{u} - \mathbf{v}) d\mathbf{u} d\mathbf{v}$ (for a stationary process $\gamma(\mathbf{u}, \mathbf{v})$ is a function of only $\|\mathbf{u} - \mathbf{v}\|$), and γ can be estimated by solving an inverse problem for γ:

$$z_i z_j = \int_{A_i} \int_{A_j} \hat{\gamma}(\mathbf{u} - \mathbf{v}) d\mathbf{u} d\mathbf{v}, \quad i, j = 1, \ldots, N. \quad (4)$$

If the A_i's are regularly spaced, the appropriate $z_i z_j$ could be averaged, reducing the number of equations. In the general case, it is not clear how to reduce the total number of $N * (N + 1)/2$ equations.

For an isotropic and stationary process $\gamma(\mathbf{u}, \mathbf{v}) = \gamma(\|\mathbf{u} - \mathbf{v}\|)$, where $\|\cdot\|$ is Euclidean distance. Equation (4) can be rewritten as:

$$
\begin{aligned}
z_i z_j &= \int_{A_i} \int_{A_j} \gamma(\mathbf{u} - \mathbf{v}) d\mathbf{u} d\mathbf{v} \\
&= \int_0^\infty W_{A_i A_j}(l) \gamma(l) dl, \quad (5)
\end{aligned}
$$

where $l = \|\mathbf{u} - \mathbf{v}\|$ and

$$W_{A_i A_j}(l) = \int_{\mathbf{u}, \mathbf{v}: \mathbf{u} \in A_i, \mathbf{v} \in A_j, \|u-v\|=l} d\mathbf{u} d\mathbf{v}$$

There are two important difficulties trying to solve the Equations (5) with respect to γ:

1. Large number of equations. For a single disease in our example we have 48 spatial regions (states) and 372 intervals in time. When solving equations in space only we have $48 * 49/2 = 1176$ different equations. When solving in space-time we get $48 * 372 * (48 * 372 + 1)/2 = 159,427,296$ different equations.

2. Representation of a potential solution $\hat{\gamma}$. We are trying to find a covariance function which restricts our search domain to a space of positive definite functions.

To address the first problem we develop a fast algorithm to compute the weight function $W_{A_i A_j}(l)$. An efficient algorithm to calculate the kernels $W_{A_i A_j}$ when the regions A_i are piecewise polygons in R^2 or prisms with a piecewise polygon base in R^3 is described in Mockus (1994).

To address the second problem we may use a parametric family of covariance functions (i.e., $\gamma(x,y) = \sigma e^{\alpha \|x-y\|}$). Another approach is to re-express the covariance function in terms of some simpler function. A general form of an isotropic correlation function ($d \geq 2$) (Equation 6) can be found in, e.g., Matèrn (1986).

$$r(l) = G(0) + \int_0^\infty \phi_Y(ls) dG(s), \qquad (6)$$

$$\phi_Y(l) = \frac{d-2}{2}! \left(\frac{2}{l}\right)^{\frac{d-2}{2}} J_{\frac{d-2}{2}}(l)$$

where G is a distribution function. The isotropic correlation function is a scale mixture of Bessel functions and the estimation of such function could be done by estimating the function G appearing in Equation (6).

Reformulating problem (5) in terms of the function G we get

$$z_i z_j' = \int_0^\infty W_{A_i A_j}(l) \gamma(l) dl$$
$$= \int_0^\infty W_{A_i A_j}(l) \sigma \left(G(0) + \int_0^\infty \phi_Y(ls) dG(s)\right) dl$$
$$= \int_0^\infty Q_{A_i A_j}(s) dG(s)$$

where S_{A_i} is the area of A_i, $Q_{A_i A_j}(s) = \sigma\left(\int_0^\infty W_{A_i A_j}(l) \phi_Y(ls) dl + S_{A_i} S_{A_j} G(0)\right)$, and $\sigma = \gamma(0)$.

We can look for the weighted least squares solution to Equations (7) and (5).

$$\hat{\gamma} = \arg\min_\gamma \sum_{i,j} C_{ij} \left(z_i z_j - \int W_{ij}(l) \gamma(l) dl\right)^2, \quad (7)$$

$$\hat{G} = \arg\min_G \sum_{i,j} C_{ij} \left(z_i z_j - \int Q_{ij}(s) dG(s)\right)^2. \quad (8)$$

where C_{ij} are weights. The C_{ij}s should be selected to be inversely proportional to the variance of $z_i z_j$. We can approximate C_{ij}s by $\left(\frac{1}{S_{A_i} S_{A_j}}\right)^2$. In practice we can perform the least squares iteratively; first find $\hat{\gamma}$ using $C_{ij} = \left(\frac{1}{S_{A_i} S_{A_j}}\right)^2$, then set $C_{ij} = \frac{1}{(\int W_{ij}(l) \gamma(l) dl)^2}$ and perform least squares

again. The C_{ij} are updated in this way until $\hat{\gamma}$ stops changing from iteration to iteration. In practice, having the weights $C_{ij} = \left(\frac{1}{S_{A_i} S_{A_j}}\right)^2$ (as opposed to having $C_{ij} = 1$) provided more reliable convergence of the numerical optimization methods in finding the optimal solution $\hat{\gamma}$.

4.2 Example

4.2.1 Generated data

To illustrate the behavior of the covariances between regions we first consider an example with simulated data and the covariance function $\gamma(t) = \sigma \int_0^\infty J_0(ts) dG(s)$, where $G(s) = \begin{cases} 0 & if \quad s < 10 \\ 1 & if \quad s > 10 \end{cases}$ and J_0 is Bessel function of the first kind. The 48 regions $A_1, \ldots, A_{48}$ partition the map of United States (Figure 2). The map (see Becker and Wilks (1991)) is given in Albers equal area projection (see Deetz and Adams (1921)) coordinates which were rescaled so that the area of the map would be equal to 1. All distances in this paper refer to the distances obtained in that way. For comparison, on our scale the distance of 3000 miles is approximately 1.

Figure 2: The 48 regions

Figure 3 shows exact covariances between all possible pairs of regions. The covariances $\int_{u \in A_i} \int_{v \in A_j} \gamma(u-v) du dv$ are plotted on the Y axis, while the average distances between regions are plotted on the X axis; average distance d_{ij} between regions A_i, A_j is given by:

$$d_{ij} = \frac{\int_0^\infty W_{ij}(l) l \, dl}{\int_0^\infty W_{ij}(l) dl}.$$

There are $48(48+1)/2 = 1176$ points in the plot. The points approximately follow the line of $\gamma(l)$.

Figure 4 shows observed covariances between all possible pairs of regions. The values $z_i = \int_{A_i} f(x) dx$ for each region were generated. Then

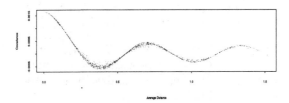

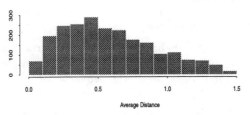

Figure 3: The plot of covariances versus average distance (top) and the histogram of average distances (bottom).

averages $X_{ij}^{10} = \sum_{k=1}^{10} z_i^k z_j^k$ of products $z_i z_j$ over ten realizations of the process f are plotted against the average distance between the regions A_i, A_j (Figure 4) with LOWESS estimate (see Cleveland (1979)). The grayscale image (dark corresponds to high density) is a scaled two-dimensional histogram of the scatterplot of X_{ij}^{10} versus d_{ij}. The scaling is inverse proportional to the density of average distances, i.e., the value n_{xy} of the scaled histogram at bin x, y is:

$$n_{xy} = N_{xy}/M_x \qquad (9)$$

where N_{xy} is the number of (X_{ij}^{10}, d_{ij}) pairs falling into bin x, y, and M_x the number of points d_{ij} falling into bin x. To increase the smoothness of the aforementioned histogram we generated 100 realizations of X_{ij} for a total of $48(48+1)/2 \times 100 = 117600$ points.

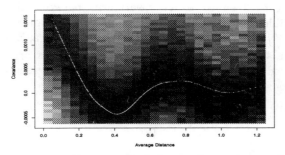

Figure 4: Histogram of observed covariances (background color) and LOWESS smoother (large white dots).

We then estimated parameters (α, σ) from 100

Quantile	50	25	75
α	10.5	8.2	11.5
σ	1.76	1.1	2.55

Table 2: Quantiles of estimated parameters. The parameters used to generate the data were $\alpha = 10$, $\sigma = 1$

realizations of the process f with the covariance function $\gamma(t) = \sigma \int_0^\infty J_0(\alpha s) ds$ by solving Equation (8) with $C_{ij} = 1$. The results are in Table 2.

4.2.2 Disease data

The incidence rate process of disease data often has seasonal and longer time trends (see Figure 1) that need to be removed before estimation of the covariance function. In addition, each state has its own disease reporting mechanism.

Let z_{ij} be the reported incidence rates in state (geographic/administrative/political region) i for time interval j. To reduce the noise we aggregated the data into four week periods, so j corresponds to one month. To remove time and state trends we used median polish to fit state effects s_i and time effects t_j. The residuals

$$\eta_{ij} = z_{ij} - s_i - t_j$$

for any particular state did not have an obvious trend.

Covariances $E(\eta_{i,j}, \eta_{k,j+l})$ for different lags l are shown in Figure 5. The residuals were normalized since we are only interested in the shape of the covariance function, not the scale. After normalization the scatterplot (average distance on x-axis and covariance on y-axis) was smoothed by LOWESS and shown in Figure 5. The products $E(\eta_{i,j}, \eta_{i,j+l})$ were not included in the scatterplot since those correlations are always non-negative. We see that while both types of hepatitis have similar covariances indicating no clear mid-range (relative to the geographic size of the United States) isotropic correlations, the Tuberculosis data indicates short range correlations decreasing with geographic distance.

The estimated covariance function confirms the the same point. We tried fitting parametric covariance function models. We estimated parameters (α, σ) for the covariance function $\gamma(t) = \sigma e^{\alpha s}$ by solving Equation (8) with $C_{ij} = 1$. The data was averages of products $\eta_{i,j} \eta_{k,j+l}$ over all time intervals. The estimated exponent parameters for

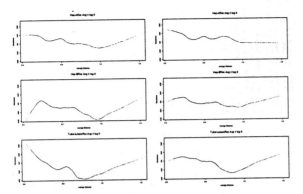

Figure 5: LOWESS smoother of observed isotropic covariances. Lag zero is on the left and lag 6 is on the right. All plots have the same scale with zero on the vertical axis in the midle of the scale. Hepatitis A is on top, Tuberculosis - at the bottom.

various lags are in Table 3. Note, that region-to-itself $(\eta_{i,j}\eta_{i,j+l})$ covariances were used in estimation, while the scatterplots smoothed by LOWESS (in Figure 5) did not include those covariances. The results confirm the plots in Figure 5, namely, that spatial isotropic correlations decrease fastest for Tuberculosis, followed by Hepatitis A (which is also an infectious disease), but for Hepatitis B which spreads by a different mechanism (by contaminated blood products) correlations behave differently. The spatial correlations decrease with the time lag. The exponential covariance model was not the best model since at geographic distances equal to 2000 miles all estimated covariances seem to be negative.

Disease	Lag 0	Lag 3	Lag 6
Hepatitis A	-23	-1.5	-15
Hepatitis B	-6.0	-6.1	-3.5
Tuberculosis	-190	-169	-30

Table 3: Estimated exponential parameter for various diseases and lags

5   Summary

We designed and implemented a set of methods to estimate dependence structure of a spatial-temporal stochastic process when the available data is represented in the form of integrals of the considered process. We applied our techniques to simulated and real data. We concentrate on spatial (two-dimensional) isotropic covariance function and we estimate arbitrary parametric form of the covariance function. We describe and illustrate a way to re-express covariance function as a nonnegative function for the purpose of numeric non-parametric estimation.

Our estimation method is based on solution to a system of integral equations (4) via numeric optimization. In simulated data we were able to recover the original covariance function. In real data concerning infectious diseases we obtained spatial correlations that reasonably describe different mechanisms by which each disease spreads.

Our main purpose was to obtain a quantitative estimate of the variability and smoothness for the predictor of a stochastic process.

6   References

Becker, R. A. and Wilks, A. R. (1991). Maps in S. *AT&T Bell Laboratories Statistics Research Report*

Cleveland, W. S. (1979). Robust locally weighted regression and smoothing scatterplots. *Journal of the American Statistical Association* **74**, 829-836.

Cressie, N. (1985). Fitting Variogram Models by Weighted Least Squares. *Mathematical Geology.* **17**: 563:568.

Deetz and Adams (1921). Elements of Map Projection. *USGS Special Publication* No. 68, GPO.

Dyn, N. and Wahba, G. (1982). On the Estimation of Functions of Several Variables from Aggregated Data. *SIAM J. Math. Anal.*, **13**:1 134-152.

Eddy, W.F. and Mockus, A. (1994). An Example of the Estimation and Display of a Smoothly Varying Function of Time and Space - The Incidence of the Disease Mumps. *Journal of the American Society for Information Science.* **45**(9): 686-693.

Eddy, W.F. and Mockus, A. (1995). Discovering, Describing, and Understanding Spatial-Temporal Patterns of Disease Using Dynamic Graphics. Systems.

Marshall, R.J. and Mardia, K.V. (1985). Minimum Norm Quadratic Estimation of Components of Spatial Covariances. *Mathematical Geology.* **17**:517-525.

Matèrn, B. (1986). *Spatial Variation* (2nd ed.). Lecture Notes in Statistics. **36**. Berlin. Springer-Verlag.

Mockus A. (1994). Predicting a Space-Time Process from Aggregate Data Exemplified by the Animation of Mumps Disease. PhD Thesis. Department of Statistics, Carnegie Mellon University.

Tobler, W.R. (1979). Smooth Pycnophylactic Interpolation for Geographic Regions. *Journl of the American Statistical Association.* **74** 367: 519-536.

ANALYSIS OF THE EFFECTS OF INDUCTION THERAPY WITH STEROID FREE MAINTENANCE IMMUNOSUPPRESSION ON A GROUP OF HTX PATIENTS

Flora O. Ayeni, Medtronic, Inc.

Medtronic, Inc., 7000 Central Avenue, N.E. T278, Minneapolis, Minnesota 55432

Key Words: Transplant, Induction Therapy, Immunosuppression, Hypercholesterolemia

This paper focuses on the graphical analysis of the data obtained from a study of two groups of transplant patients. The first group consists of high risk transplant (HTX) patients (n=24) selected to receive OKT3 induction therapy for 10 to 14 days with steroid free immunosuppression (SFMI) after six weeks. This group was compared to a second group of transplant patients (n=36) that received triple drug immunosuppression (TDI) during the same time interval. Total serum cholesterols were measured at pre-transplant, and at six months and one year post-transplants. The data obtained was analyzed using both statistical and graphical methods. Several plots were used to display and compare patient's cholesterol levels for both groups. The results clearly show that OKT3 induction with steroid free immunosuppresion (SFMI) after heart transplant prevents the development of hypercholesterolemia and diabetes in treated patients, with no increase in mortality or incidence of rejection.

INTRODUCTION

A control chart is a simple chart with statistically determined upper control limit (UCL) and lower control limit (LCL) drawn on either side of the overall average. The sample averages are plotted on the chart (X-bar chart) to determine whether any of the points falls outside of the control limits or forms "unnatural" patterns. If either of these happen, the process is said to be out of control. The points within the control limits result from the natural variation in the process. These are known as common cause variations. They are part of the system. However, points outside the control limits come from a special cause that is not part of the way the process normally operates. This special cause must be eliminated.

In this paper, we evaluated two groups of heart transplant patients. The first group consisted of high risk transplant patients (n=24) who were selected to receive OKT3 induction therapy for 10 to 14 days with steroid free maintenance immunosuppression (SFMI) after six weeks. This group was compared to a group of transplant patients (n=36) that received triple drug

immunosuppression (TDI) during the same time interval. Total serum cholesterols were measured at pre-transplant, and at six months and one year post-transplants. The data obtained were analyzed using graphical methods. Several graphs were used to display and analyze patient's cholesterol levels for both groups.

GRAPHICAL ANALYSIS

Several plots were made to analyze the effects of induction therapy (OKT3) or triple drug immunosuppression (TDI) on each patient under study. Figure 1 shows a relatively stable pre-transplant cholesterol level for patients in both OKT3 and TDI. However, after 6 months of post-transplant, there is an upward trend in the cholesterol levels of patients in the group as compared to patients in the OKT3 (Figure 2). At one year post-transplant, the distinction between the cholesterol levels of TDI and OKT3 patients can be easily seen (Figure 3).

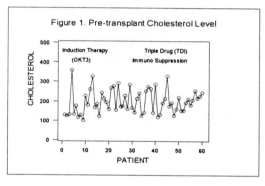

Figure 1. Pre-transplant Cholesterol Level

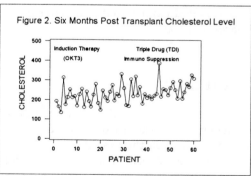

Figure 2. Six Months Post Transplant Cholesterol Level

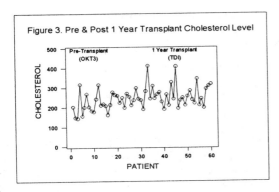

Figure 3. Pre & Post 1 Year Transplant Cholesterol Level

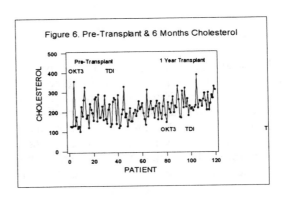

Figure 6. Pre-Transplant & 6 Months Cholesterol

Figure 4 shows the relationship between pre-transplant, 6 months, and one year patients. The graphs show an upward trend of patient's cholesterol level from pre-transplant to 6 months and one year for groups OKT3 and TDI.

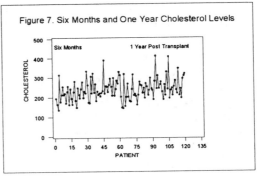

Figure 7. Six Months and One Year Cholesterol Levels

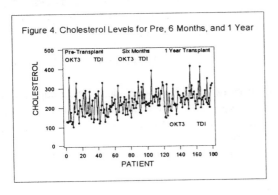

Figure 4. Cholesterol Levels for Pre, 6 Months, and 1 Year

CONTROL CHART ANALYSIS

The control chart method does not require assumptions of normality, equivalence of the variances or constancy of the cause systems. For these reasons, it differs from t-test and the analysis of variance. This paper presents two methods of control charts:

RANGE CHART (R-CHART)

The R-chart is used in this paper to study the variation in cholesterol level from patient to patient for both OKT3 & TDI groups. The R-chart is in control, an indication that the variation in the cholesterol level over the 3 study periods is consistent from patient-to-patient. The overall standard deviation is 47.34 for both groups (Figure 8).

Several pair-wise graphical comparisons were provided: (a) pre-transplant and 1 year (Figure 5), (b) pre-transplant and 6 months (Figure 6), and (c) 6 months and 1 year post-transplant (Figure 7).

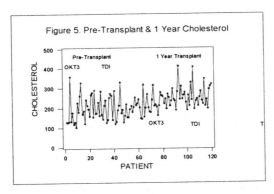

Figure 5. Pre-Transplant & 1 Year Cholesterol

Figure 5 shows that the one year post-transplant is significantly different from pre-transplant. This is because the one year post-transplant cholesterol level is trending upward while the pre-transplant shows no trend. Similar conclusion can be made for 6 months post-transplant in Figure 6. The trend is fairly similar between 6 months and one year post-transplants.

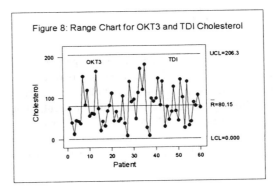

Figure 8: Range Chart for OKT3 and TDI Cholesterol

AVERAGE CHART (X-BAR CHART):

The X-bar chart is used to monitor the average cholesterol level for each patient over the three study periods. In Figure 9, the X-bar chart is not in control. This is due to the unusual high average cholesterol level for patients #4, #33, and #45. The major objective now is to identify the major causes of these high cholesterol levels for these three patients. After identifying these causes, one should work to eliminate them.

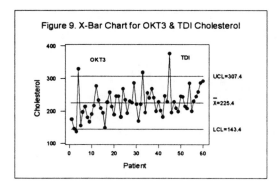

Figure 9. X-Bar Chart for OKT3 & TDI Cholesterol

R-CHARTS FOR WITHIN GROUP STUDY

In addition to the above combined charts for OKT3 and TDI groups, the R and X-bar charts are provided for each group separately. These enable us to understand the variation in the cholesterol level within each group of patient and compare the result from the two groups with each other.

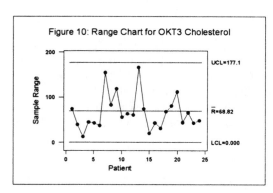

Figure 10: Range Chart for OKT3 Cholesterol

Figure 10 shows the range chart for patients that received the OKT3 drug. The range chart is in control indicating that the variation in the cholesterol level is consistent (homogeneous) from patient-to-patient. The standard deviation is 40.65 for OKT3 patients. Similarly, a range chart is provided in Figure 11 for the patients in the TDI group. This chart is also in

control, indicating that the variation in cholesterol level is consistent from patient-to-patient.

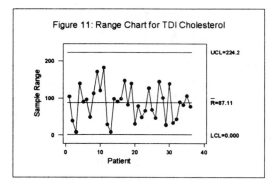

Figure 11: Range Chart for TDI Cholesterol

The standard deviation is 51.45 for TDI patients. In conclusion, this result shows that the variation in the cholesterol level from patient-to-patient is smaller (Std.Dev.=40.65) for OKT3 patients than for TDI patients (Std. Dev.= 51.45).

X-BAR CHARTS FOR WITHIN GROUP STUDY

In Figures 12 and 13, we consider X-bar charts for each group separately. These charts allow us to monitor separately the average cholesterol level over the three study periods. The results obtained here can be used to compare the average cholesterol level for the two groups.

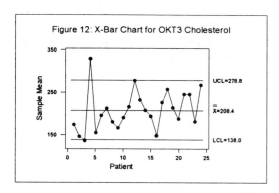

Figure 12: X-Bar Chart for OKT3 Cholesterol

The X-bar chart for the patients with OKT3 drug is not in control. This is because patient #4 has an unusual average cholesterol level not consistent with other OKT3 patient. This patient needs to be looked at very carefully to identify the causes that made the average cholesterol level special from the rest of the patients within the OKT3 group.

The overall average cholesterol level for OKT3 patient is 208.4. The X-bar chart for the patients with triple drug (TDI) is provided in Figure 13. This chart is also not in control. In this group, patient #21 has an average cholesterol level that is outside the control limits, indicating that this patient's cholesterol level is not consistent with other patients on TDI treatment.

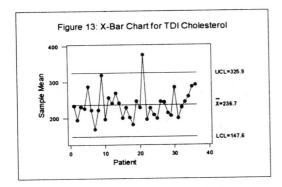

Figure 13: X-Bar Chart for TDI Cholesterol

The overall average cholesterol level for TDI patients is 236.7. In conclusion, patients on TDI treatment have higher overall average cholesterol level than patients on OKT3 treatment. This shows that the average cholesterol level (236.7) of TDI treatment is significantly different from the average cholesterol level (208.4) of OKT3 treatment.

CONCLUSIONS

Advances in computer technology have made possible that medical data can be displayed as control charts. The potential of statistical process control charts in medicine is unlimited. Among the advantages of using control charts in medical practice includes balancing the amount of drugs used in the management of patient care and the automatic indication of body functions changes.

The control chart agrees with a 't' test or analysis of variance if the X-bar charts are in control and the R-charts are in control meaning the variability has been constant and equal. In the cholesterol study, the range chart is in control, however the X-bar chart is not in control.

In conclusion, we found patients managed by OKT3 induction with SFMI after transplant have less variation (Std. Dev. = 40.65) in their total serum cholesterol level than their counterpart who received triple drug therapy (Std. Dev. = 51.45).

REFERENCES

1. Kume Hitoshi (1985), "Statistical methods for quality improvement", 3A Corporation, Japan.

2. Besterfield Dale (1979), "Quality Control", Prentice Hall.

GRAPHICAL PRESENTATION IN CROSS-OVER TRIALS

C. Gordon Law, Edward F. C. Pun, Pfizer Central Research
C. Gordon Law, Pfizer Central Research, Eastern Point Rd., Groton, CT 06340

Key Words: Exploratory plots, Carry-over, Intra-subject variation

1. INTRODUCTION

In crossover trials, subjects are randomized to receive a sequence of two or more treatments. As responses to different treatments are collected from the same subject, between subject variation can be eliminated. If used properly, a crossover design usually requires fewer subjects than a parallel design.

In this manuscript, current graphical approaches for presenting the data in a 2x2 crossover trial will be reviewed. The classic data set analyzed by Grizzle (1965) is used for illustration. We propose an idea which can easily be programmed. The idea is extended to higher order crossover trails. An example for presenting data from a two-sequence, four-period trial is given.

2. REVIEW OF CURRENT GRAPHICAL PRESENTATION FOR 2x2 CROSSOVER TRIAL

Let Y_{ijk} be the response of the i-th subject in Period j (j=1,2) of Sequence k (k=1,2). Each subject provides a pair of observations for two treatments over two periods. Some within-subject quantities for the i-th subject are defined below:

Period Sum = $Y_{i1k} + Y_{i2k}$,
Period Difference = $Y_{i1k} - Y_{i2k}$, and

Treatment Difference = $\begin{cases} Y_{i11} - Y_{i21} & \text{if Sequence 1} \\ Y_{i22} - Y_{i12} & \text{if Sequence 2} \end{cases}$

These quantities can be used to explore the carry-over effect, the treatment effect and the period effect, respectively (Jones and Kenward, 89). Several graphical methods had been suggested to present the data in the above setting.

Subject Profile Plot

Figure 1 displays the Grizzle's data by treatment sequence. The profile of each subject is described as their responses over Periods 1 and 2 are· lined up. It is difficult to explore the treatment or the carry-over effects in this framework.

Groups-by-Periods Plot

The mean response by group (treatment sequence) and period of the Grizzle's data are plotted in Figure 2. This figure is constructed by calculating the mean of each data cluster in Figure 1 and overlaying the summary in Sequence 2 to Sequence 1. A detailed description of how to interpret different patterns in this type of plot can be found in Jones and Kenward (89).

Sum Difference Plot

Figure 3 presents the period sum and the period difference of the Grizzle's data. In this figure, one can explore the carry-over effect and the treatment effect by comparing the data clusters along the horizontal axis and the vertical axis, respectively. The use of the convex hull can assist in visualizing the data clusters by group.

Sliding-Square Plot

This plot was proposed by Pontius and Schantz (94). The center portion in Figure 4 presents the paired response with different treatment. The boxplots at the upper and the right margins show the treatment variation. The boxplots at the lower right corner display the distribution of the period sum multiplied by $1/\sqrt{2}$. The variation of the period difference (also multiplied by $1/\sqrt{2}$) is summarized by the boxplots at the lower left corner. A differentiating feature of this plot is that it takes different variations associated with the 2x2 crossover trial into account and can be used to explore the distributional assumptions.

A comparison of the above graphical approaches is given in Table 1. Although the sliding-square plot has advantage over the other approaches, it is

FIGURE 1: Subject Profile Plot for Grizzle's Data

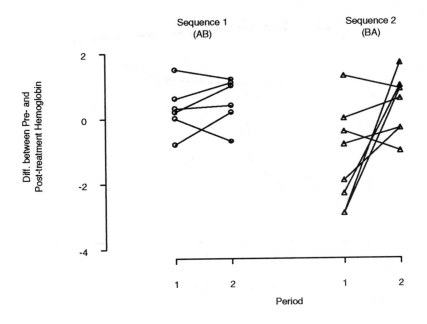

FIGURE 2: Groups-by-Periods Plot for Grizzle's Data

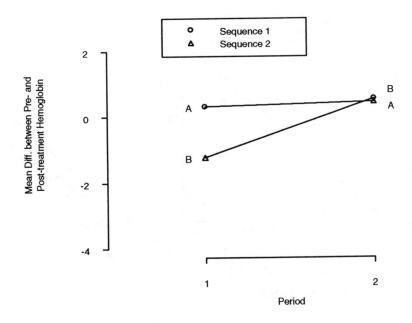

FIGURE 3: Sum Difference Plot for Grizzle's Data

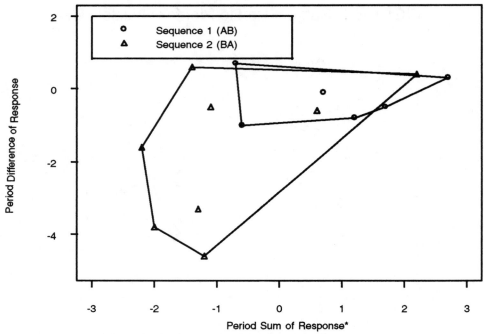

Period Difference of Response

Sequence 1 (AB)
Sequence 2 (BA)

Period Sum of Response*

* Difference between pre- and post-treatment Hemoglobin

FIGURE 4: Sliding-Square Plot for Grizzle's Data

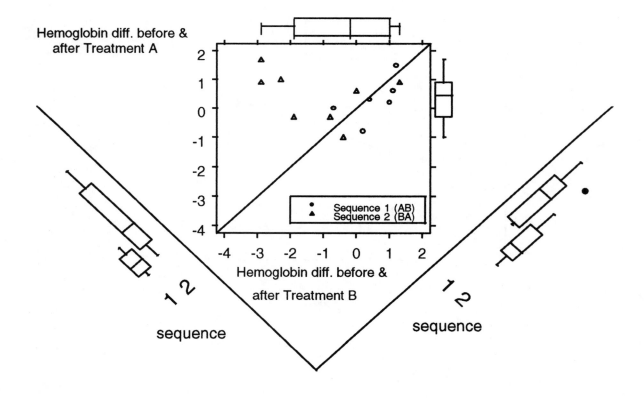

Hemoglobin diff. before &
after Treatment A

Hemoglobin diff. before &
after Treatment B

Sequence 1 (AB)
Sequence 2 (BA)

sequence

sequence

difficult to implement in practice. A large programming effort is needed to construct the plot. Moreover, the boxplots at the lower left and right corners have no labeling on the corresponding axes. Thus, the information contained in the boxplots cannot be quantified.

3. MODIFICATION OF SLIDING-SQUARE PLOT

From a programming viewpoint, we propose a dual display plot. The graphical region of the plot is divided into two sections. One section is used for displaying the raw data. The other section is used to explore the variation of some within-subject quantities. The plot can be programmed easily using S-PLUS, as illustrated in Figure 5. The left half of the figure is the same as the center portion of Figure 4. The boxplot for the period sum, period difference, and treatment difference are presented at the right half of the figure. As Grizzle's data set is a classic example to illustrate carry-over effect, it is not surprising to find a significant difference in the period sum comparison.

4. GRAPHICAL PRESENTATIONS FOR HIGHER ORDER CROSSOVER TRIALS

Higher order crossover trials extend the usual 2x2 crossover trials in either the number of sequences or the number of periods. With optimal crossover designs, subjects may experience a sequence of identical treatment in two successive periods. Independent estimates for within subject variability can be derived. Even with the presence of carry-over effect, the designs still allow us to estimate the treatment effect.

For higher order crossover trials with the number of sequences greater than 2, the previous idea can still be implemented by using a different symbol for each sequence and deriving some within-subject contrasts for exploratory analysis. When the number of periods is greater than 2, the raw data can be presented with multiple period windows.

For illustration, an example in Chow and Liu (92, P.279) is used. The example covers a data set under the two-sequence, four-period design

$(j=1,2,3,4)$. In Figure 6, three windows are introduced to display the subject profile. As all the axes have the same scale, the profile of each subject can easily be traced across all periods. The following within-subject contrasts,

$$U_{ik} = Y_{i1k} + 2Y_{i2k} - 2Y_{i3k} - Y_{i4k} \,,$$
$$V_{ik} = Y_{i1k} - Y_{i2k} - Y_{i3k} + Y_{i4k} \,, \text{ and}$$
$$W_{ik1} = |Y_{i1k} - Y_{i4k}| \,, \quad W_{ik2} = |Y_{i2k} - Y_{i3k}| \,,$$

are selected to explore the carry-over effect, the treatment effect, and the within subject variation, respectively. The medians of the U contrasts of Sequences 1 and 2 approximate each other. Although the variation in Sequence 1 is higher than in Sequence 2, there is probably no carryover effect. In the absence of unequal carry-over effects, the comparison of the V contrasts suggests the possibility of a treatment difference. The boxplots of the W contrasts show that Treatment B incurred smaller intra-subject variation than Treatment A. Detailed analysis of the data can be found in Chow and Liu (92).

REFERENCES

1. Chow, SC and Liu, JP (1992), *Design and Analysis of Bioavailability and Bioequivalence Studies*, New York: Marcel Dekker.

2. Grizzle, JE (1965), "The Two-Period Change-Over Design and Its Use in Clinical Trials," *Biometrics* 21, 467-480.

3. Jones, B and Kenward, MG (1989), *Design and Analysis of Cross-over Trials*, New York: Chapman & Hall.

4. Pontius, JS and Schantz, RM (1994), "Graphical Analysis of a Two-Period Crossover Design," *The American Statistician*, 48, 249-253.

5. Rosenbaum, PR (1989), "Exploratory Plots for Paired Data," *The American Statistician*, 43, 108-109.

6. Statistical Sciences, Inc., *S-PLUS for Windows*, Version 3.2 (1994), Seattle, WA: Author.

Table 1: Comparison of Current Graphical Methods for 2x2 Crossover Study

	GRAPHICAL METHOD			
OBJECTIVE	Subject Profile	Groups by Periods	Sum Difference	Sliding Square
Present Data	Yes	*No*	*No*	Yes
Explore Treatment Effect	Yes	Yes	Yes	Yes
Explore Period Effect	Yes	Yes	*No*	Yes
Explore Carry-over Effect	*No*	Yes	Yes	Yes
Explore Variability	*No*	*No*	Yes	Yes

FIGURE 5: Dual Display Plot for Grizzle's Data

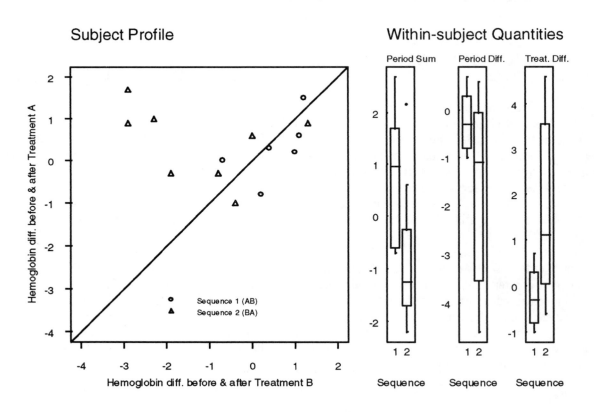

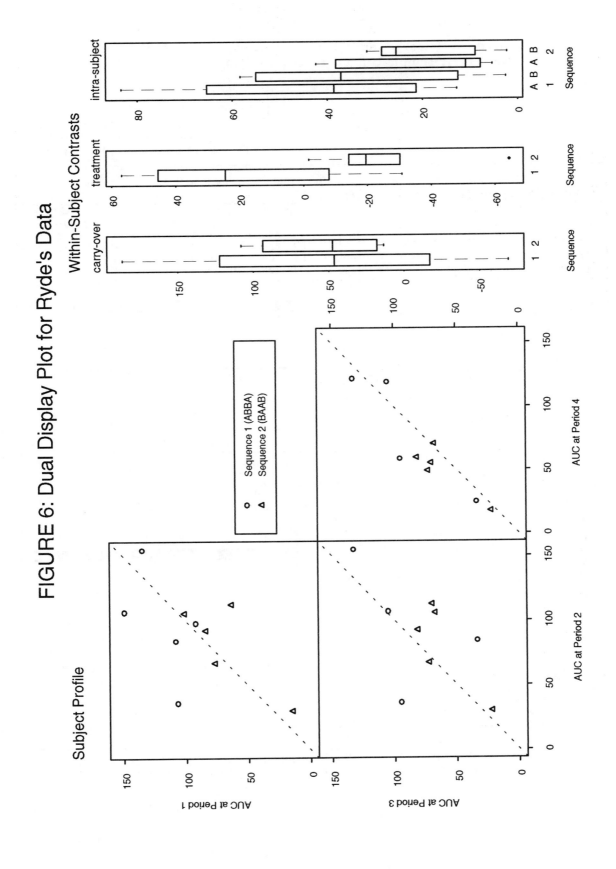

FIGURE 6: Dual Display Plot for Ryde's Data

DESIGN OF A SPREADSHEET INTERFACE FOR S

Richard M. Heiberger and Magnus Mengelbier, Temple University
Richard M. Heiberger, Temple University, Department of Statistics, Philadelphia, PA

Key Words: Splus

Abstract

We have designed and constructed an interactive spreadsheet interface to S (Becker, Chambers, and Wilks 1988) that maintains the complete power and generality of the S language. At user level, the interface behaves like the popular spreadsheet programs available for personal computers: one or more S data objects (matrices or three-way arrays) are displayed on the screen. The user graphically (with mouse or cursor motion) identifies a cell for review, and possible updating, of its contents. The user can graphically identify one of the `spread.frame`'s associated macros, containing an arbitrary S expression, and then either update the macro's definition or execute it.

The current version, available on statlib (send spread from S), uses the classes and methods technology introduced in S Version 3 (Chambers and Hastie, 1992). The spreadsheet is designed in a modular fashion with device-specific methods for the display and updating of spreadsheet objects. We include methods for two devices: the generic S graphics device and a character based device using the emacs environment. The goal is an S Version 4 implementation that takes full advantage of the event-driven readers to be made available.

We discuss the design decisions mandated by the joint constraints of providing complete spreadsheet capability along with the full power of S, and the portability considerations for working with several different screen handling technologies.

1 Introduction

A spreadsheet interface to S, as in Figure 1, can simplify several situations. We use it for simultaneous data collection, display, and summarization. It is useful for data handling and cleaning operations, expecially when working with multiple linked files. It can provide interactive multiple views of a single data file. It is also useful for statisticians working in environments where other people expect data to be spreadsheet compatible.

A spreadsheet in S differs in several important ways from an ordinary spreadsheet. Data is entered directly as an S object, hence the data conversion

Figure 1. Cell Update Forces Re-evaluation

a. The entire `spread.frame` is displayed in a screen buffer. The user identifies one cell for updating, usually by moving a pointing device or cursor. The formula associated with the cell is displayed in a window.

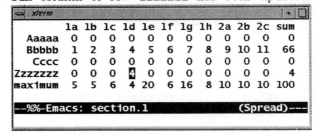

b. The user changes the cell formula.

c. The system evaluates the new cell formula, and then reevaluates the entire `spread.frame`. The sum column of row Zzzzzzz has been updated.

step can be avoided. Any S function can be associated with the entire spreadsheet or any of its cells. The full analytic and display power of S is available.

We provide spreadsheet behavior by introducing a new class of S objects, the `spread.frame`, and several supporting methods. A `spread.frame` is an object of class `"spread"`, derived from the `"array"` class, with one or more attributes of mode `"expression"`. The default structure maintains four types of expressions: `"expr"`, `"before"`, `"after"`, and `"macro"`. The array `"expr"` and the vectors `"before"` and `"after"` are evaluated every time the `spread.frame` object is updated. When any cell of the `spread.frame` is changed, any other cell whose value is dependent on that cell is automatically updated.

Any valid S expression may be associated with the `spread.frame`. Since the `spread.frame` object is an `array` with additional attributes, it can be used in ordinary calculations.

2 Spreadsheet Kernel

2.1 `spread.frame` Object

The `spread.frame` object inherits from an ordinary matrix, data.frame, or three-way array. Any pre-existing object can be coerced to a `spread.frame` by the S statement `x <- as.spread(x)`. The `spread.frame` object `x` has the following four new attributes (any or all of which may be empty) in which executable expressions are stored.

`attr(x,"expr")` An array of expressions with shape identical to that of the data array within the `spread.frame x` itself. Each element of `"expr"` may be an arbitrary S expression that evaluates to a scalar. The expression may be a constant, or be dependent on other cells of the `spread.frame`, or be dependent on other objects in the `search()` path. The value of each expression is assigned to the matching cell of the data array. The entire `"expr"` is evaluated every time any cell of the `spread.frame` object is modified.

`attr(x,"before")` A vector of arbitrary S expressions. The entire vector is evaluated every time any cell of the `spread.frame` is modified and before the cell-by-cell expressions in the `"expr"` array. A typical application would be to modify/update one or more values of the `spread.frame` in response to changes in an associated variable .

`attr(x,"after")` A vector of arbitrary S expressions. The entire vector is evaluated every time any cell of the `spread.frame` is modified and after the cell-by-cell expressions in `"expr"`. A typical application would be to summarize information in the updated `spread.frame`. It could also be used to update another S object or a graphical display.

`attr(x,"macro")` A vector of arbitrary S expressions. Any individual expression is evaluated on request from the user.

2.2 Infrastructure: "spread" Methods

The fundamental method that drives the process is subscripted assignment $[\leftarrow$. The simplest case, scalar assignment to a single cell, is

```
"[←.spread" ←
    function(object, ..., value)
{
    x ← object[,]
    e ← attr(object,"expr")
    e[[...]] ← substitute(value)
    x[[...]] ← eval(e[[...]])
    attributes(x) ← attributes(object)
    attr(x,"expr") ← e
    update(x)
}
```

The `"expr"` attribute is removed from the object and updated with the *formula* of the assignment, not the *value*. Then the formula for the updated cell is evaluated in the frame of the `"[←.spread"` function. This has the implication that other cells in the `spread.frame` must be referenced by the name `x[,]`. Finally, the `update()` function is called to evaluate the `"before"` expressions, cycle through the entire set of `"expr"` expressions, and then evaluate the `"after"` expressions.

The behavior is illustrated in Figure 2.

2.3 Design Decisions

Our initial design of the `"spread"` class included just two types of expressions, the `"expr"` and `"macro"`. These two types are all we need to obtain the behavior of the popular PC spreadsheets. We added the other two types for two reasons: execution efficiency and simplicity.

Simple matrix summaries, such as a set of row sums, are efficient in S when executed as a single statement. The identical arithmetic executed on a cell-by-cell basis does not use the generic array structure of S and hence bears a heavy performance penalty. For example, the gradesheet application in Section 4.1 has essentially one summary operation, computing a set of row sums. In an array language such as S, doing this with a single statement in the `"after"` vector makes more sense and is faster than the cell-by-cell method with individual summations for each cell in the `"sum"` column.

Having both `"before"` and `"after"` vectors allows symmetry in linking `spread.frames` to other S objects or external information resources such as databases and real-time sensor arrays.

Figure 2. Simple `spread.frame` with Formula Assigned to Cells

```
> tmp                              # ordinary matrix
      1 2 3 4 sum
Aaaaa 0 0 0 0   0
Bbbbb 0 0 0 0   0
 Cccc 0 0 0 0   0

> x <- as.spread(tmp)             # turn it into a spread.frame
> x[1,"sum"] <- sum(x[1,-5])      # assign formula to cells
> expr(x)[[1,"sum"]]              # formula is stored
sum(x[1, -5])

> x[1,1:4] <- c(10, 20, 30, 40)   # assign values
> x
      1  2  3  4 sum
Aaaaa 10 20 30 40 100             # sum column uses formula
Bbbbb  0  0  0  0   0
 Cccc  0  0  0  0   0

> names(attributes(x))
[1] "dim"       "dimnames" "expr"       "macro"
[5] "before"    "after"    "class"
```

2.4 Multiple `spread.frames`

The current implementation supports multiple `spread.frames` and multiple views of the same `spread.frame` through a single S session. We have introduced several objects to simplify the switching between `spread.frames`. The `spread.frame` kernel assumes the `spread.frame` is an S object named `x` as described in Section 2.2. The switching task is essentially a mechanism for keeping track of the user's name for the active `spread.frame`. We do this with a foreground function `fg()` and two frame-0 variables `.Active` and `.Active.buffer`. `.Active` contains the name of the active `spread.frame` object and `.Active.buffer` contains the name of the current two-dimensional slice. When the name of a `spread.frame`, say `"section1"`, is assigned to `.Active`, its value is duplicated in the S variable `x` and the screen display is updated. Upon completion of the update, the resulting object is assigned back into the original name stored in `.Active`.

The object `.Registry` is a data structure used to keep track of the `spread.frames` in the current `.Data`. In the current version `.Registry` is itself a `spread.frame` whose macro names are the names of the `spread.frames`. Our intent is to make `.Registry` a more general nested list. Its function will be to keep track of `spread.frame`-specific format information. Currently, the display appearance is that of the default `print()` command. Eventually, the display will be formatted according to the information stored in `.Registry`. The object `.Registry` is retained across login sessions.

All this occurs transparent to the user and, ideally, the user will work entirely through spreadsheet style interfaces, or similar buffers, and will never need to interact directly with an S session.

3 Device Handler Structure

The simplicity of the spreadsheet is that the user graphically identifies a cell for review, and possible updating, of its formula. For discussion purposes we will use the terminology of *placing the cursor on the cell*. When the cursor is placed on a cell entry, the cell formula is displayed for update. The underlying structure of a `spread.frame` is specified by the definition of the class `"spread"`. Details of identifying the cell and of displaying a formula for update are device specific. The idea of defining a series of `spread.frame` devices is similar to S supporting a series of graphics devices.

The current emacs device works by constructing S command sequences and appending them to the end of the S session buffer, essentially the same as if a user had typed it. Future development will make the behavior more like that of a smart graphics device with a spreadsheet mode. The end result will be a spreadsheet style interface which is a natural extension of the parser environment. An event-based interface is under construction to be used with Version 4 of S.

3.1 emacs device: Spread mode

We are working with Emacs 19.28. In this implementation of the emacs interface, the emacs environment is in charge and S is a subordinate process. The details are based on the comint (compiler interface) mode introduced with Emacs 19. We have defined a new read-only major mode, Spread mode, which permits cursor movement (cursor arrows and page-up and page-down commands), the ENTER key, and several ^C-prefix keys. On X-windows terminals Spread mode accepts mouse clicks.

The effect of the ENTER key is to:

1. place in the minibuffer an S assignment statement that will re-create the current cell value from the current cell expression.

2. wait for a minibuffer entry.

Once the user has modified the minibuffer entry with a replacement expression, emacs will

3. print the modified statement in the *S* buffer, tell S to execute it, and finally print the updated spread.frame to the spreadsheet buffer.

The emacs interface is built on the comint mode. The user must first create a .Data directory in the appropriate directory. Then Spread mode is entered by loading an emacs lisp file sprd-int.el. The lisp file creates a new buffer *Splus* and runs Splus in it. If .Registry already exists in .Data it is displayed in Spread mode, otherwise a new .Registry is created and displayed. Existing spread.frames are displayed by typing "f" or [mouse-2] on the name of the spread.frame in .Registry. Once a spread.frame has been displayed, all further review or change can be made entirely in Spread mode. Cursor or mouse motion is used to select a cell. The ENTER key or [mouse-3] is used to open a cell's formula for display or revision. Macros associated with the spread.frame can be displayed, changed, or executed by typing or clicking.

The mechanics of the interface are simple in principle. Each key or mouse action causes emacs to generate, type, and execute one or more S commands in the *Splus* buffer. Execution is initiated by the emacs (comint-send-input) function. Emacs waits for the S command to complete with the (accept-process-output) function. The emacs (comint-output-filter) function detects any errors in the excution of the S command. When an S error is detected in one of a series of S commands, all further generated S commands are cancelled and the display shifts to the *Splus* buffer.

All generated command sequences end with an S command to print the revised spread.frame to a temporary file, followed by an emacs command to read that file and display it in Spread mode.

The *Splus* buffer is always available for the user to enter S commands manually. Two common uses are for the initial creation of a spread.frame and to investigate S execution errors.

A major advantage of the emacs interface is the lack of dependence on workstation graphics display technology. It will will work well on any character-based terminal for which a termcap definition exists. In particular, the emacs device interface can be used over slow telephone connections.

3.2 Graphics Device

The graphics device interface is still in the existence demonstration phase. The screen is divided with the split.screen() functions into small pieces, one per cell of the spread.frame. The entire spread.frame is plotted with the text() command. spread.frame cells are identified by a control function based on the locator() function. Once the cell has been located with the mouse, the function will:

1. place in the S window an S assignment statement that will re-create the current cell value from the current cell expression.

2. wait for an S entry.

Once the user has modified the constructed assignment statement, using workstation-specific screen editing, and entered the modified statement, the control function will

3. execute the modified statement and print the updated spread.frame to the graphics device.

Potential advantages of the graphics device include the ability to place one or more graphs in other windows of the same graphics device in which the spread.frame itself is displayed.

3.3 Future Development

Version 4 of S has facilities for event monitoring. The spread.frames will be able to send their updating information directly to a FIFO running in the background of the S session. Currently this information is sent to the **tty** in which S is running.

We will extend the infrastructure to include higher-dimensional arrays and, more generally, arbitrary subsets of data objects. We will support Import/Export of popular spreadsheet file formats. We will have SQL capability and be able to display arbitrary views of database formats.

We have outlined a tcl/tk interface that will work on both Windows and Unix version of S-Plus. The current emacs interface requires a Unix host. Although the graphics device interface works on both versions, it is still very rudimentary.

4   Examples

4.1   Gradesheet

I frequently collect data on several variables for a group of students and need to construct a summary of the results (for example, a grade sheet showing the points earned by each student on each exam question, and the total score). I construct a **spread.frame** with the student names as the row identifier, the question numbers as the column identifier. I construct an **"after"** expression

```
after(x)["sum"] <- expression(
   x[, 12] <- apply(x[, -12], 1, sum) )
```

that sums the values for the questions and places them in the **"sum"** column.

The **spread.frame** after one student's scores have been entered appears in Figure 1a. Note that I have placed the maximum scores for each sub-question in the **spread.frame**. This functions both as an aid when entering the data and as a permanent record when the filled out **spread.frame** is printed for my class notes book.

I find it easiest to record grades by entering a student's entire set of question scores at once. I therefore created a sub-class **"grade"** that inherits from **"spread"**. The **"grade"** class has one defining characteristic, when the cursor is placed on the student's name, the entire row except for the **"sum"** column is opened for data entry. (Normally, when a row label is entered, the whole row is opened.) For example, when I enter student **Bbbbb** by clicking on the **Bbbbb** cell in Figure 1a, the entry panel displays

```
x["Bbbbb",-12] <-
     c(0,0,0,0,0,0,0,0,0,0,0)
```

I modify this to the appropriate values from the marked exam:

```
x["Cccc",-12] <-
     c(1,2,3,4,5,6,7,8,9,10,11)
```

Figure 3.   **after()** Changes Graphs as Point is Moved. Changing the value of a single point automatically changes the graph of the data. On the left graph, the point **x[11,"y"] = 0.379**. On the right graph, the point **x[11,"y"] = 4.379**.

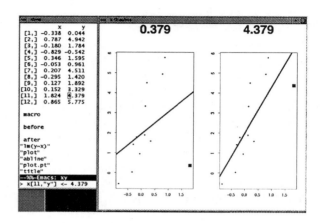

On entry, the **spread.frame** updates itself and calculates the sum for that student. It would be easy to update the stem-and-leaf display or other summary graphic of the sums, as each individual student is added to the **spread.frame**.

4.2   Linking S Objects

The names of the macros of the **spread.frame .Registry** (described in section 2.4) are the names of other **spread.frames** (for example, **section.1**, **section.2**, **section.3** for an instructor teaching three sections of a course). Execution of one of the macros causes the named **spread.frame** to be displayed. The screen display of **.Registry** is therefore a menu into the other **spread.frames**.

4.3   Outliers in Regression

In Figure 3 we have created a **spread.frame** that contains two columns of data, **X** and **Y**. We have created an **"after"** expression that regresses $Y \sim X$, plots the data with emphasis on the first point, and then draws the regression line. Every time a data point is changed the plot is redrawn. The application is to illustrate the effect of changing the value of one point on the regression line. We create an array of plots and then systematically change the value of the first point in the data. The resulting series of plots shows that a data point on the edge of the **X** range has a stronger effect on the slope than a data point in the center of the range.

Figure 4. Three-way `spread.frame`: Gradesheet with **sum** and **course** macros.

a. *sec3[,,"midterm"]*

```
          1  2  3  4 sum
Aaaaa    19 20 18 21  78
Bbbbb    24 24 19 21  88
Cccc     13 46 83  0 142
maximum  25 30 20 25 100
```

b. *sec3[,,"final"]*

```
          1  2  3  4 sum
Aaaaa    20 18 30 15  83
Bbbbb     0  0  0  0   0
Cccc      0  0  0  0   0
maximum  25 25 35 15 100
```

c. *sec3[,"sum",]*

```
         midterm final course
Aaaaa        78     83    161
Bbbbb        88      0     88
Cccc        142      0    142
maximum     100    100    200
```

4.4 Multiple Views of a `spread.frame`

The interface can display multiple coordinated views of a single `spread.frame`. Figure 4 shows three views of a three-way grade sheet. Two slices correspond to two exams and a third slice, orthogonal to the first two, shows the course summary. The `spread.frame` has two **after** macros to caclulate the row sums for each exam and the course sums across exams. On an X-terminal, each slice might appear in its own window by taking advantage of emacs 19's ability to control multiple windows in a single emacs session. On a text-based terminal, the different buffers appear in the same emacs window.

The following commands are generated from each graphical interaction:

1. The commands

```
> attach(
+    '/disk5/rmh/spread/sprd3d/.Data')
> emacs.start('/tmp/spra00792')
```

are issued when the `spread.frame` interface is started. They attach the S library con-

taining the `spread.frame` functions and call `emacs.start()` with the name of the directory where temporary files will be stored.

2. The command

```
> print.find.emacs('sec3[,,"final"]',
+    update.Registry=F)
```

is generated when the user places the cursor on the `sec3[,,"final"]` line in `.Registry`. This line prints all known views of the `sec3` object to temporary files and then reads them into emacs buffers.

3. The command

```
> emacs.cell('sec3[,,"final"]',2,1,1)
```

is generated when the user places the cursor on a row label in the `sec3[,,"final"]` buffer. The S function places in the minibuffer an S assignment statement that will re-create the current cell value from the current cell expression.

4. The commands

```
> x[ "Aaaaa",-5,"final" ] <-
+              c(20, 18, 30, 15)
> invisible(assign(.Active, x))
```

are generated when the user edits the minibuffer. The execution of the revised assignment updates the entire `spread.frame` and causes all views of the updated `spread.frame` to be displayed in their emacs buffers. The assignment of the value of the working `spread.frame` x to the `spread.frame` named in `.Active` keeps S's idea of the `spread.frame`'s value synchronized with the user's idea.

4.5 Your Favorite Spreadsheet Application

Any spreadsheet should be able to be written in the S setting. This gives the double advantage of the familiar and powerful S analysis and graphics operations coupled with the ease of data entry from working in spreadsheet mode.

References

Becker, R. A., J. M. Chambers, and A. R. Wilks (1988), *The New S Language: A Programming Environment for Data Analysis and Graphics*, Wadsworth, Monterey, CA.

Chambers, J. M., and T. J. Hastie (1992), *Statistical Models in S*, Wadsworth, Monterey, CA.

SMALL-COLLEGE FACULTY SALARY MODELS

Ann Russey Cannon, Cornell College   Douglas M. Andrews, Wittenberg University
Ann Russey Cannon, Box 8148 Cornell College, 600 First St. W., Mt. Vernon IA 52314

1   Introduction

In taking part in the 1995 ASA graphics section data analysis exposition, we chose to concentrate on how faculty are paid at small liberal arts colleges, since both of us work at such schools. We were motivated at least in part by a desire to check whether the pay at our own institutions is in line with the pay at other small colleges. That Ann is on the Faculty Salary Subcommittee at Cornell is another motivation for this investigation. And, although we were interested primarily in small colleges, we could not resist comparing the Type IIB schools with the larger Type I universities. The data sets used for this analysis were supplied by U.S. News and World Report, and the AAUP.

2   Response Variable Choice

The AAUP dataset contains data on both salary and total compensation, reported for each rank and as an overall average. There is, of course, a very strong association ($r = 0.99$) between average salary and average compensation. Hence we decided to focus on just one or the other; we chose salary because the definition was clear. It was not clear that all schools had the same definition of compensation.

It is no surprise that full professors are paid more than associates, who in turn are paid more than assistants. But moreover, the schools that pay well overall, pay well consistently across the ranks – and vice versa. In fact, the correlations between overall average salary and average salary for full, associate, and assistant professors are 0.97, 0.94, and 0.93, respectively. These trends hold for schools of all AAUP types, and for public and private schools alike. Hence we limited our analysis to modeling the average salary for all ranks combined as the response variable.

3   Explanatory Variables

We categorized the remaining variables in the AAUP and U.S. News datasets by the following general factors that we felt would be relevant in predicting the average faculty salary:

- General Classification
  - AAUP type
  - Private/Public
- Selectivity
  - quartiles and mean for ACT and SAT
  - percent of students in top 10% and 25% of their high school class
  - percent of applicants accepted
  - percent of accepted students enrolled
- Faculty Caliber
  - percent of faculty with PhD
  - percent of faculty with terminal degree
- Cost
  - tuition (in-state and out-of-state)
  - room and board
  - fees, books, personal spending
- Size
  - number of full- and part-time undergraduates
  - number of faculty at each rank and total number of faculty
  - number of applicants, acceptances, and matriculants
- Miscellaneous
  - student/faculty ratio
  - percent of alumni who donate
  - graduation rate
  - percent of faculty at each rank
  - instructional expenditure per student

Although the groups were determined in a somewhat *ad hoc* manner we suspect that they represent general factors that would help explain faculty salary structure – at least at schools like ours. Certainly there may be better ways to measure these general factors than the variables we have available. For example, we might like to have data on admissions criteria for assessing selectivity, or perhaps some measure of professional or research activity for assessing faculty caliber. With this in mind we merely used this categorization of the explanatory variables to organize our model selection and to help interpret why some variables were more important than others in predicting faculty salary.

4 Overall Salary Trends

Most academics would probably guess that Type I schools pay much better than Type IIB schools, and some might guess that publicly funded schools pay better than privately funded schools. But few might be aware of the interesting interaction between the AAUP type and the source of the funding. As can be seen in the boxplots in figure 1, the private Type I schools pay more than the public Type I schools, whereas for the Type IIB schools the publicly funded ones have a slightly higher average salary. Standard ANOVA corroborates the significance of this interaction. It is worth noting that even though the difference between private and public salaries is quite a bit larger for Type I schools than for Type IIB schools, this is not enough to pull the overall private average higher than the public average, as only about 30% of the 176 Type I schools are private whereas about 84% of the 599 Type IIB schools are private.

5 Model for Type IIB Schools

5.1 All Type IIB Institutions

Although there was a significant interaction between AAUP type and the public/private distinction, the median salary was comparable between the public and private Type IIB's (with a difference of only about $2000). Hence we set out to develop a model for faculty salary at *all* Type IIB schools. From the outset we did not believe this would be the best model for either group since, despite the similar median, the spread among average salaries was quite different between the public schools and the private schools.

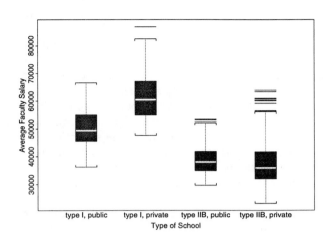

Figure 1: Boxplot of average salaries by school type

Not surprisingly, the variables in each of our subsets are associated with one another – to a greater or lesser degree. So we suspected that we would need no more than one or two variables from each subset, as any others would be redundant for the purposes of our model. This was indeed the case, as will be described below for each subset of the explanatory variables.

5.1.1 Selectivity

One of our first hypotheses was that the more elite schools, which attract brighter students, can afford to pay their faculty well. The 14 variables in the US-News dataset that we put into this category afforded us a handy way to test this suspicion.

The first way to look at selectivity is through the scores on standardized tests. Of the 10 variables measuring test scores, it turns out that the lower quartiles (for Math SAT, Verbal SAT, and ACT alike) are more useful than the means and upper quartiles. For example, consider the relationship with salary for these three measures on the ACT. The correlations (with salary) for the lower quartile, mean, and upper quartile are 0.72, 0.65, and 0.63, respectively; hence the R^2 and F-statistic from simple linear regression would be highest for the lower quartile. From the ANOVA, it became apparent that each of the three would be a decent predictor by itself, but that only the lower quartile contributed significantly after inclusion of the other two. A similar trend holds for the Math and Verbal SAT scores.

Finally, we decided to include only the lower quartile of the Math SAT in our final model, as no other test score lower quartile contributed significantly to our model after this variable was entered.

Why is the lower quartile a bit more useful in predicting faculty salaries? Admissions offices are fond of bragging about the cream of each new class. But measures of the middle of the pack probably are a more helpful measure of overall student caliber. Moreover, measures of the weaker students are probably a better indicator of the school's overall academic standards. Those schools which must admit a higher proportion of stereotypically weaker full-tuition students to balance the budget each year will tend to be more hard-pressed to keep faculty salaries competitive. Surely such students will drag down the mean score as well. But their effect on the lower quartile seems to be more pronounced.

As an alternative way to assess selectivity, we generated the proportion of applicants accepted and the proportion of accepted students enrolled from the raw counts of applications, acceptances, and enrollments reported in the U.S. News dataset. We suspected that the schools which were more selective in this sense would pay their professors more, and this was indeed the case. The proportion of applicants accepted, as an indicator of selectivity also offers more in the way of predictive power for faculty salaries over and above the lower quartile of the Math SAT.

5.1.2 Faculty Caliber

Perhaps the most fair predictor of faculty salary would be some measure of the overall quality of the faculty. Unfortunately, the only such measures available in this dataset are the percentages of faculty with Ph.D.'s and with terminal degrees. The former was a bit more predictive, and the latter offered no significant contribution over and above the former.

5.1.3 Cost

Among the variables pertaining to the cost of each institution, the one most strongly associated with salary was tuition. And as with the other subsets of variables, no other factor contributed significantly after tuition was entered in the model.

There is certainly some logic to tuition's importance, as faculty compensation is usually one of the largest components in any small school's operating budget. The additional fees vary somewhat arbitrarily by school, and the ambiguity inherent in estimating book costs and personal spending surely adds

plenty of unwanted variation to those variables. Finally, the room and board costs probably pertain more to the efficiency of the food service and the quality of the dorms than to faculty salaries.

5.1.4 Size

Even among the Type IIB institutions the larger schools pay their faculty more, as was clear from the predictive power of the number of full-time undergraduates. As another measure of size we looked at the number of faculty at each rank. The number of full professors was helpful in the prediction, but as discussed in the next section, the *proportion* of full professors is a better predictor. Another possible measure of size was the number of applicants. This factor was significant, but did not increase the R^2 value appreciably, so, in the interest of a leaner model, we did not use this factor in the end.

5.1.5 Miscellaneous

From the reported *counts* of faculty at each rank we generated the *proportions* at each rank. And, as anticipated, the proportion of full professors had a strong (positive) association with overall salary – even stronger than the (negative) association between salary and the proportion of assistant professors. These proportions pertain not to the size but to the age of the faculty and hence the amount that must be paid to the faculty. But regardless, the strength of its relationship with salary warrants its inclusion in our model.

Because the bulk of the instructional expenditure (at Type IIB schools that emphasize undergraduate education) is on faculty salaries rather than equipment, we were not surprised that the instructional expenditure per student was also a worthy predictor.

5.1.6 Final Model

So for all Type IIB schools, we were left with the following explanatory variables in our model:

Variable	Coef.
intercept	204.1
lower quartile of Math SAT	.1458
proportion of applicants accepted	-85.11
percent of faculty with PhD	.8728
out-of-state tuition	.0017
number of full-time undergraduates	.0224
proportion full professors	71.67
instructional expenditure per student	.0060

It should be noted that where we have used the term "percent," the values are given in whole numbers (e.g. 60) and where we have used the term proportion the values used are in decimal form (e.g. 0.60).

Type III sums of squares from the analysis of variance suggest that each of these factors bears information not already given by any other variable in the model. And, except for the fact that we measure selectivity in two different ways, we needed only one variable from each of our subsets. But is this model appropriate for both public and private institutions?

5.2 Public and Private Type IIB Institutions

For the most part, we believe that the above model is adequate for all Type IIB schools. But for some of our variables, the relationship with average salary appears to differ for the public and private schools. As an example, consider figure 2 which is a plot of salary and out-of-state tuition. There is indeed a clear positive linear relationship for the private schools, but little if any relationship for the public schools. Such differences led us to develop separate models for the two kinds of institutions. For the private schools, we ultimately selected the following predictors:

Variable	Coef.
intercept	93.76
lower quartile of Math SAT	.2907
percent of students in top 10% of class	.3515
proportion of full professors	77.22
percent of faculty with PhD	.7580
out-of-state tuition	.0044
instructional expenditure per student	.0029

That most of these variables coincide with the variables in the combined model should be no surprise, as about 500 of the 600 Type IIB's are privately funded. For the publicly funded Type IIB's, however, the final model had only three significant, nonredundant factors:

Variable	Coef.
intercept	358.4
proportion of applicants accepted	-109.0
proportion of full professors	150.79
room and board	.0214

That there are only three worthwhile predictors might be due in part to the fact that there are far fewer public schools, so that the trends and relationships are not as apparent, and the effects of extreme or unusual schools are much more pronounced. Regardless, the R^2 for the private IIB model is 0.74, compared to only 0.52 for the model for public IIB's.

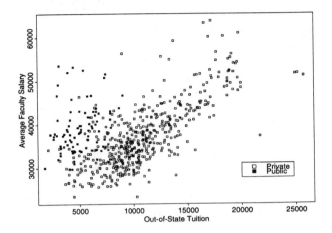

Figure 2: Scatterplot for Type IIB schools of average faculty salary and tuition broken down by public vs. private

We also note that, aside from having fewer variables, the public model looks very different from both the overall Type IIB model and the private model in terms of the actual variables used. For instance, tuition is gone and is replaced by room and board. Is this because public schools have more control over setting room and board than tuition? Gone from this model also are all measures from standardized tests. This last finding was very surprising to us and we have not been able to adequately explain it.

6 Prediction and Comparison

Naturally, we thought to use our models to assess the pay at our own schools. Cornell reports an overall average salary of $46,800. The combined Type IIB model predicts this to be $40,859 with a standard error of $434; the model for private Type IIB's predicts $43,421 with a standard error of $366. (These standard errors are for the mean response for all schools with Cornell's model characteristics.) Wittenberg reported $43,800; the combined and private Type IIB models predicted $43,401 and $44,666 with standard errors of $476 and $454, respectively. These standard errors are fairly representative of all standard errors for predictions from all the schools. That the variability of our predictions is smaller for the model for private schools alone should not be a surprise.

7 Surprisingly Bad Predictors

There were several variables given in this dataset that we thought would be quite helpful in our model for faculty salaries, but which turned out to be no help at all:

- *graduation rate* Surely the schools which are more selective (and presumably pay their faculty better) admit better students, who then are more able to survive until graduation. But the relationship with overall salary was negligible. Upon further reflection, because such schools are more rigorous academically, it is not clear whether the graduation rate would be positively or negatively related to the general notion of selectivity, let alone faculty salary.

- *student/faculty ratio* We anticipated that the more selective schools can afford a (proportinately) larger faculty and can also afford to pay them more. No such relationship was apparent. Perhaps those schools that are putting lots of resources toward improving faculty salaries are doing so by hiring fewer new professors, and those schools that are hiring more must suffer smaller salary increases.

- *percent of donating alumni* Here we reasoned that such donations can be a significant portion of the incoming funds for the schools with more loyal alums. Such schools might also have larger endowments and hence can afford to spend more of their operating budgets on faculty salaries. But this was not borne out by the data.

8 Summary

While we were aware of the salary differences between Type I and Type IIB schools, we had not anticipated the strong interaction between AAUP type and the public/private distinction. That salary at the Type IIB private schools could be modeled more precisely than at public schools is probably a consequence of the fact that we had far fewer public schools than private schools in our data set.

Finally, to study faculty salaries more carefully, we believe we would have found useful some community information (e.g. the size of the community,

the cost of living), the religious affiliation (if any), the teaching workload required, and the amount of the financial aid budget (perhaps as a proportion of the overall budget) for each school.

9 Acknowledgements

We would like to thank Tim Hesterberg for his comments on an earlier version of this paper which led to a revised and (hopefully) better paper.

STATISTICAL PROBLEMS, PITFALLS AND PARADOXES: ANALYSIS OF U.S. COLLEGE DATA

John Lawrence, The Ohio State University
Dept. of Statistics, Room 141 Cockins Hall, 1958 Neil Avenue, Columbus, Ohio 43210

There are many problems that result from the collection and electronic entry of large data sets. For the data used in the *US News* rankings of colleges [Morse (1994)], particular colleges may have incorrectly answered certain questions due to clerical error (for example, transposed adjacent digits) or may have interpreted a specific question differently than others. There are more problems with this particular data set than there is generally because this data is self reported by colleges and they have a vested interest in putting the best possible "spin" on the data that they report. J. W. Gilley, the president of Marshall University, compares the way that some schools misrepresent data to bolster their rankings to Faust selling his soul to the devil [Gilley (1992)]. He continues with his attack on these rankings by saying that it is impossible to fairly rank schools with self reported data-"Though *US News* has cleaned up its research formula, its foundation is still built on quicksand. For no matter how well-designed the formula, if the input is questionable, the resulting rankings can not be valid". A front page story in the *Wall Street Journal* [Stecklow (1995)] documents numerous differences between data reported to ratings guides and data that colleges reported to debt-rating agencies and investors. In nearly every case, the data reported to debt-rating agencies, which by law must be factual, was less favorable to the colleges. It's similar to price mistakes at supermarkets- we wouldn't mind the mistakes so much if they were in our favor half the time, but when they're consistently in the store's favor then there is a problem. I had intended to use this data set to make a presentation about the graphical depiction of Simpson's paradox. However, I quickly discovered that there were many problems with the data set. Since no analysis should be undertaken on data which is known to be faulty, I decided to try to characterize some techniques for detecting problems with data and methods of correcting these problems. With this data set, there were other sources for data which could be used to check the accuracy of the data reported here, but this is not always the case. It is always possible to look at scatterplots and histograms of the variables to see if the data is self consistent or within admissible bounds and this will be the first technique we consider. Next, some of the reported variables will be compared with other sources.

This is a plot of "percent of faculty with terminal degree" versus "percent of faculty with Ph. D.'s":

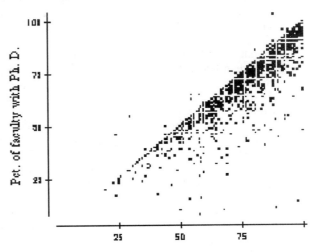

Pct. of faculty with terminal degree

There are apparently some schools where there are more faculty with Ph. D.'s than there are faculty with terminal degrees. Also, there is one school where the faculty are really trying to give a hundred and ten percent (103 percent of the faculty have Ph. D.'s).

Three schools reported more new students enrolled than were accepted (in fact, one reported 4385 more new students enrolled than were admitted). One school had a graduation rate of 118 percent. One school's "First quartile- Math SAT" was 10 points lower than their "Third quartile- Math SAT" score. Nine schools were incorrectly reported as public or private.

The scatterplot of "Average Combined SAT score" versus the sum of "Average Math SAT score" and "Average Verbal SAT score" shows that there were some errors in the reporting of these scores:

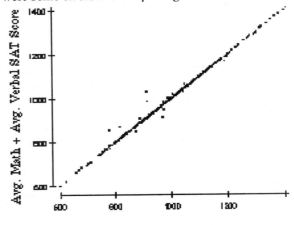

Avg. Combined SAT Score

US News used these scores in part to calculate the "Student selectivity" score. Florida Southern University reported an average Math SAT score of 505, an Average Verbal SAT score of 478 and an Average Combined SAT score of 912. Since it's not possible for any student to only take one part of the exam, the sum of the averages of the Math and Verbal scores should be the Average Combined score unless medians were used as the average. However, individual SAT scores are always divisible by 10, so the median would always be a multiple of 5. The only reasonable conclusion is that there was an error in reporting or calculating these scores. It makes a difference in the *US News* rankings since Florida Southern and Carson-Newman College were tied for tenth place in the Southern Regional Liberal Arts Colleges category. Specific formulas were not available so that the overall scores could be recalculated, but it is entirely possible that if Florida Southern's Combined SAT score were 70 points higher than the reported value, there would no longer be a tie for tenth place and Florida Southern could move up to ninth place. This is important because only the top ten schools in the category are listed at all.

Here is a scatterplot of "First quartile- Math SAT" versus "Third quartile- Math SAT":

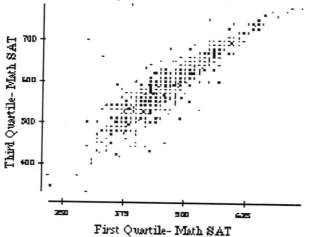

What is unusual here is that a few schools have a remarkably large difference between the first and third quartile. There are at least two different explanations for this large difference- either there was a mistake made in reporting the scores or these schools had a very small number of students who took the SAT. The latter explanation brings up an important problem in analyzing this data. Some schools require the standardized exams and some do not. It may not be appropriate to treat the scores reported from schools that require the exams the same as the scores reported from those that do not. In other words, if a school only had 10 people take the

SAT and they report an average combined SAT score of 700, that may not be indicative of the standardized test taking ability of that college's population.

The *Wall Street Journal* article discusses several discrepancies that were found by comparing the data reported to *US News* with data reported to Moody's. For example, Harvard University gave Moody's a range of SAT scores with a midpoint of 1385. They did not give *US News* an overall range, but they did give a range for Math scores and Verbal scores individually. The sum of the two midpoints was 1400 and this is the score that was used by *US News* for Harvard's average SAT score. It's not possible to determine if Harvard intentionally did not report a midrange for the total SAT score to *US News*, and it is not my intention to insinuate that they did. I would like to point out that in this same article, a former admissions officers from a different college recounted how officials "huddled at a meeting that could only be described as a strategy session on how to cheat on the survey". The article also describes how scores are commonly inflated by leaving out scores from groups (for example, part-time students) that would otherwise deflate the college's average.

Another variable that is often "massaged" by colleges is graduation rate. There is typically a difference between the graduation rate reported to *US News* - where higher rates are more favorable to the college, and the NCAA- where lower scores are more favorable to the college since the college-wide graduation rate is compared to the graduation rate of student athletes. The figure below is a box plot of the difference between the graduation rates reported to *US News* by 50 of the top schools and the graduation rates for the same year found in The College Handbook:

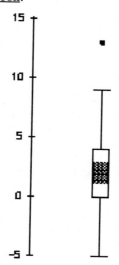

Difference in graduation rates

Wright (1992) and Webster (1992) describe problems with the intrinsic nature of some of the variables used in the rankings. For example, academic reputation is determined by a survey that is distributed to college officials. In *US News'* first ranking of colleges, this was the only variable used. Now, it only counts for twenty-five percent of the ranking, but it is still very influential and obviously subject to bias. The survey respondents are asked to make a list of the top quartile of all schools in the category. For instance, in the National Universities category, there are 229 schools, so they would list the top 57 schools in the category. Harvard, MIT and Stanford tied for appearing on the most people's lists, while Princeton, Yale, Chicago, Johns Hopkins, and UC Berkeley tied for making the second most people's lists. Most people would be hard pressed to make a list of even twenty universities in the entire world which have a better reputation than Princeton or UC Berkeley. The president of the University of Rochester has said that college presidents were sure to vote for three institutions: their employers, their alma maters, and their doctoral institutions [Wright (1992)].

Bibliography

The College Handbook (1995), College Entrance Examination Board, New York.

Gilley, J. W. (1992), "Faust Goes to College", *Academe*, 78, p. 9-11.

Morse, R.J., et al. (1994), "America's Best Colleges", *US News & World Report* (Cover Story, September 26), p. 86-119.

Stecklow S. (1995), "Cheat Sheets: Colleges Inflate SATs and Graduation Rates in Popular Guidebooks; Schools Say They Must Fib to *US News* and Others to Compete Effectively Moody's Requires the Truth", *Wall Street Journal* , April 5, p. 1.

Webster, D. (1992), "Rankings of Undergraduate Education in *US News* and *Money*: Are They Any Good", *Change*, March/April, v. 24 no. 2, p. 19-31.

Wright, B. (1992), "A Little Learning is a Dangerous Thing", *The College Board Review*, 163, Spring, p. 6-16.

1995 Higher Education Directory (1994), Higher Education Publicatrions, Inc., Virginia.

THE STRUCTURE OF AMERICAN COLLEGES:
UTILITY OF CUTTING-EDGE VISUALIZATION TOOLS FOR EXPANDING CLUSTER ANALYSIS

Douglas Luke, Saint Louis University
School of Public Health, 3663 Lindell Blvd., St. Louis, MO 63108

Key Words: Cluster Analysis, Graphics, Visualization

Cluster analysis refers to a family of statistical methods for identifying cases with distinctive characteristics in heterogeneous samples and combining them into homogeneous groups. It has become more widely used in recent years, especially in the health and social sciences. Cluster analysis provides an attractive alternative to more traditional linear model based statistics for many research situations. In particular, because cluster analysis can be used on small data sets, can be applied to any type of categorical or continuous variable, and uncovers rather than hides the heterogeneity of a data set, it is a very powerful tool for exploratory data analysis (Rapkin & Luke, 1993). However, cluster analysis as it is typically used, is relatively non-visual. As Tukey and others have stated, exploratory methods should, whenever possible, be visual methods (Tukey, 1977; Tufte, 1983). The purpose of this paper is to demonstrate the utility of various graphical and data visualization tools for extending and expanding cluster analysis procedures.

The data used in this cluster analysis come from two datasets provided by the American Statistical Association Statistical Graphics Section for their 1995 Data Analysis Exposition. The datasets contain information on over 1300 American colleges and universities and are from two sources: the 1995 *U.S. News and World Report's Guide to America's Best Colleges*, and the AAUP annual faculty salary report from the March-April 1994 issue of *Academe*.

A full cluster analysis is actually comprised of at least nine steps, including a) identifying cases for analysis, b) selecting, reducing and scaling variables, c) deriving proximity measures between cases, d) choosing the appropriate clustering algorithm, e) determining the number of clusters, f) interpreting cluster profiles, g) determining cluster stability, h) determining cluster validity, and i) presenting the cluster results (Lorr, 1983; Rapkin & Luke, 1993). However, for the purposes of this paper, we can think of cluster analysis as having three broad phases: a) the *cluster analysis preparation*, where decisions are made about which cases and variables to include in the analysis; b) *cluster analysis interpretation,* where the clusters are formed and described; and c) *cluster analysis validation*, where the meaningfulness of the clusters is determined. In each of these phases, certain graphical or data visualization tools can be used to help the investigator in making the appropriate decisions. In the rest of this paper, examples of these graphical tools are presented and discussed in relation to a cluster analysis of the American college data.

Cluster Analysis Preparation

Before the cluster analysis is actually carried out, it is important to decide which variables be included in the analysis. First, it is important to include enough variables so that there is substantive coverage, (i.e., the variables measure all the important substantive domains). However, too many variables leads to difficult profile interpretation as well as the chance that unimportant variables are included. More critically, a problematic situation arises when a subset of the variables included in a cluster analysis are highly intercorrelated. In such cases, the correlated variables tend to dominate the analysis and obscure the contribution of other measures (Hartigan, 1975).

In addition to rational and traditional empirical means of variable reduction, there exist various graphical techniques which can help the investigator choose the appropriate variables to include in the cluster analysis. Box-plots, stem-and-leaf diagrams and other univariate exploratory graphical methods are of course useful in identifying variables which may need to be excluded or transformed prior to clustering. However, it is necessary to utilize multivariate graphics to identify potential problems with variable intercorrelations.

A scatterplot matrix is invaluable for examining patterns of intercorrelations. Figure 1 displays a scatterplot matrix of five variables from the American college data. By examining the set of 10 (lower-diagonal) individual scatterplots, one can quickly see which variables are highly intercorrelated and which are not. For example, Figure 1 indicates that students' average ACT and SAT scores (AVGACT and AVGSAT) and the percentage of admitted freshman in the top 10% of their high school classes (PCT10) are highly intercorrelated, but average room and board costs and instate tuition (ROOMBRD and INTUIT) are not. Scatterplot matrices are also useful for identifying multivariate outliers prior to clustering. Cluster analysis will always assign every case to a cluster, therefore it is useful to examine outliers to determine whether they should be included in the analysis.

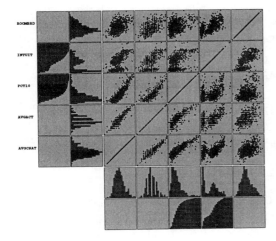

Figure 1 - Scatterplot Matrix of Five College Variables

A reduced set of seven variables was chosen to include in the cluster analysis based on intercorrelations, variability and substantive coverage. Table 1 presents the variables and their definitions. With the exception of Competitive and Scores, each variable is based on an individual variable from the American colleges dataset. Competitive is an index calculated by dividing the number of yearly student admission acceptances by the total number of yearly student applications, thus a low number indicates greater competitiveness. Scores is an index that uses either average student ACT or SAT scores, or a combined average for schools who report both figures.

Table 1. Variables Included in Cluster Analysis

Variable	Notes
• Size	Number of full time faculty at school
• Cost	Cost of instate tuition
• Salary	Average salary for all professors
• Ratio	Ratio of students to faculty at school
• Graduation	Graduation rate for all students
• Competitive	Ratio of acceptances to applications
• Scores	Standardized index based on ACT and SAT scores

Note: All variables included in cluster analysis were standardized to z-scores prior to analysis

Cluster Analysis Interpretation

972 colleges had complete data on the seven variables listed in Table 1. A hierarchical cluster analysis using Ward's clustering algorithm (Ward, 1963) was used to classify the colleges into homogeneous subgroups based on their proximities (similarities and dissimilarities). A five cluster solution was chosen based on a combined set of criteria, including the scree plot, cluster dendrogram, the cluster split-half reliabilities, and cluster interpretation. Dendrograms and scree plots are graphics commonly used in cluster analysis. (For good discussions of determining the number of clusters, including using scree plots and dendrograms, see Borgen & Barnett, 1987, Lathrop & Williams, 1990, and Rapkin & Luke, 1993.)

Table 2 presents a summary of the cluster analysis results. Each row shows the average standardized scores for a particular cluster for each of the seven variables, in addition to the cluster label and size. To interpret the results, one examines the profile patterns both across variables within a cluster, and within an individual variable across clusters. For example, the cluster *Public Universities* is made up of 209 schools that are very large (Size = 1.44), have relatively low tuition costs, have good faculty salaries (second only to *Elite Private Schools*), have relatively poor student/faculty ratios and are somewhat in the middle on graduation rate and entrance scores.

This type of summary table is the most commonly used method of presenting cluster analysis results. Although the information included in these summary tables is enough to interpret the cluster patterns, they are often confusing and inefficient, especially for analyses with large numbers of clusters or variables. There are, however, a number of graphical tools that can be used to present cluster results in a more intuitive manner. The most useful of these are grouped bar charts and parallel coordinate plots.[1]

Figure 2 presents a bar chart showing the profile patterns of the five cluster solution. Each group of bars presents the variable profile for one of the clusters. This allows a much quicker comparison of the patterns between clusters. For example, the bars for the cluster *Public Colleges* indicate that these schools have relatively low tuition costs, graduation rates and admissions scores, and relatively high student/faculty

---

[1] All of the following figures are black & white versions of the original color graphics. The use of color is an important factor in the usefulness of these graphs. Interested readers may contact the author to receive color copies of the original graphs.

Table 1. American Colleges Cluster Analysis Results

Clusters	Size	Cost	Salary	Domains			Scores
				Ratio	Graduation	Competitive	
Public Colleges (N=171)	.13	-1.07	-.43	.69	-.92	.43	-.83
Public Universities (N=209)	1.44	-.95	.74	.34	-.17	-.42	.11
Good Private Colleges (N=248)	-.41	.85	.01	-.42	.80	.28	.34
Poor Undergraduate Colleges (N=242)	-.57	.28	-.82	-.17	-.20	.30	-.39
Elite Private Schools (N = 102)	-.09	1.81	1.69	-.87	1.45	-1.43	1.76

ratios. Although bar charts are a relatively common type of graph, they have not been used frequently to present cluster analysis results.

Parallel coordinates (PC) plots are another type of graphic that are very useful for the interpretation of cluster analysis results. Parallel coordinates plots have only relatively recently been implemented in software packages such as the BMDP Diamond data visualization program.

Figure 3 is a PC plot for a random subset (N=175) of the colleges dataset. Each line running horizontally represents one case (college) from the data. Where each line crosses one of the vertical axes represents the value for that case on that particular variable. In Figure 3 the variables are the seven variables used in the cluster analysis. PC plots allow the viewer to see both the details of the data (e.g. how particular colleges compare to each other) as well as the overall patterns (e.g. the total variability of the colleges on certain variables). This follows the suggestion of Tufte, who says that

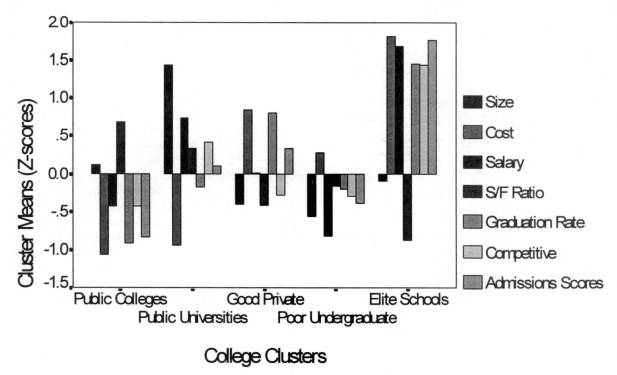

Figure 2 - Grouped Bar Chart of College Cluster Solution

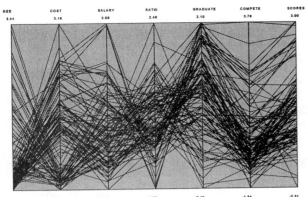

Figure 3 - PC Plot of American Colleges

good graphics should *'...reveal the data at several levels of detail, from a broad overview to the fine structure'* (Tufte, 1983, p. 13). This ability to exam data at both the micro and macro levels is further enhanced in PC plots by the ability to color code the

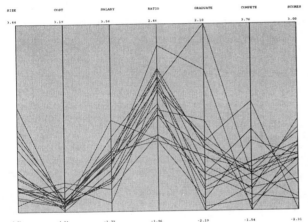

Figure 4 - PC Plot of Public Colleges Cluster

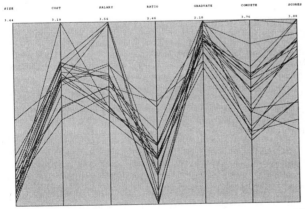

Figure 5 - PC Plot of Elite Private Schools Cluster

lines by subgroups based on a categorical variable. Although it is not clearly shown in the black & white graph in Figure, each line is color coded based on

which cluster it is in. This allows the viewer to not only observe the profile patterns by cluster (which the grouped bar chart allows), but also the variability of the individual cases *within* each cluster. This can be incredibly useful while exploring the meaning of the clusters in the interpretation phase. These inter- and intracluster patterns become clearer when we pull out the PC plots for each of the clusters one at a time (which is very easy with BMDP/Diamond). Figures 4 and 5 show the PC plots for two of the clusters, *Public Colleges* and *Elite Private Schools.* Comparing these two individual PC plots shows clearly the dramatic differences in the profile patterns of these two clusters, as well as showing the variability within the clusters. For example, there is almost no variability in the low cost of tuition (second vertical column from left) for public colleges, compared to the relatively higher amount of variability for elite private schools.

Cluster Validation

Once the final cluster solution has been determined, it is important to validate the solution by showing that cluster membership is related to other important variables in theoretically interesting ways. This is most often done by using ANOVA or discriminant function analysis using cluster membership as the primary categorical variable. Again, having graphical tools available makes the interpretation of these validation analyses more efficient. There are a number of possibilities to choose from, but probably the most useful graphical tool is the grouped box-plot.

Figures 6 and 7 show two grouped box-plots relating college cluster membership to the average amount of money spent per student and the average percentage of college alumni who donate money to their school. Both of these variables are significantly related to cluster membership, but these box-plots clearly and concisely show the differences between clusters of both the median as well as the variability of these dependent variables. For example, *Elite Private Schools* spend dramatically more money on their students, and have a wider range. Not too surprisingly, schools in the *Good Private Colleges* and *Elite Private Schools* clusters are more likely to receive money from their alumni.

Discussion

Cluster analysis is a powerful technique for both exploratory and confirmatory data analysis. The use of such cutting-edge graphical and data visualization tools as scatterplot matrices, grouped bar charts and box-plots, and parallel coordinates plots is important and

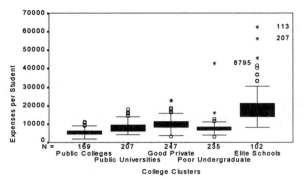

Figure 6 - Box Plot of Expenses per Student by College Cluster

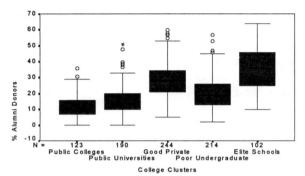

Figure 7 - Box Plot of Alumni Donor Percentage by College Cluster

useful in at least two ways. First, these graphical methods make it easier for the investigator to make important decisions about the cluster analysis including variable selection, determination of number of clusters, interpretation of clusters, and cluster solution validation. Second, these same graphical methods can be used as powerful ways to *present* the results of a cluster analysis to the intended audience. In my experience, people are more quickly and better able to 'get the story' of a cluster analysis after they have seen a grouped bar chart or PC plot than after perusing a cluster summary table.

There are no hard and fast rules about which graphical techniques to use, or how to use them. Other graphs such as foam plots (BMDP, 1995), 3-dimensional scatterplots, and jittered scatterplots (Cleveland, 1993) may be useful for cluster analysis. Or, you may find that a traditional graph presented in a slightly different format is more useful. Whenever you join the visual with the statistical, let experience, intuition and imagination be your guide.

Finally, to satisfy any reader's curiosity, Table 2 presents a few of the colleges from the dataset and the cluster that they ended up in!

Table 2. Examples of Colleges in Each Cluster

Cluster	Example Colleges
1) Public Colleges	• August College (GA)
	• Worcester State College (MA)
	• Chicago State University (IL)
2) Public Universities	• Michigan State University
	• Rutgers University (NJ)
	• University of North Carolina Chapel Hill
3) Good Private Colleges	• Saint Louis University (MO)
	• St. Olaf College (MN)
	• Scripps College (CA)
4) Poor Undergraduate Colleges	• Grambling State (LA)
	• Elmira College (NY)
	• Erskine College (SC)
5) Elite Private Schools	• Stanford University (CA)
	• Washington University (MO)
	• Oberlin College (OH)

REFERENCES

BMDP. (1995). *BMDP/DIAMOND for Windows*. Los Angeles: BMDP Statistical Software.

Borgen, F. H., & Barnett, D.C. (1987). Applying cluster analysis in counseling psychology research. *Journal of Counseling Psychology, 34,* 456-468.

Cleveland, W.S. (1993). *Visualizing data*. Murray Hill, NJ: AT&T Bell Laboratories.

Hartigan, J. (1975). *Clustering algorithms*. New York: Wiley.

Lathrop, R.G., & Williams, J.E. (1990). The validity of the inverse scree test for cluster analysis. *Educational and Psychological Measurement, 50,* 325-330.

Lorr, M. (1983). *Cluster analysis for social scientists*. Washington, DC: Jossey-Bass.

Morse, R.J., et al. (Sept. 26, 1994). America's best colleges. *U.S. News & World Report,* 86-119.

Rapkin, B.D., & Luke, D.A. (1993). Cluster analysis in community research: Epistemology and practice. *American Journal of Community Psychology, 21,* 247-277.

Tukey, J.W. (1977). *Exploratory data analysis*. Reading, MA: Addison-Wesley.

Tufte, E.R. (1983). *The visual display of quantitative information*. Cheshire, CN: Graphics Press.

Ward, J.H. (1963). Hierarchical grouping to optimize an objective function. *Journal of the American Statistical Association, 58,* 236-244.

EXPLAINING TUITION AT COLLEGES IN MASSACHUSETTS: A GRAPHICS APPROACH

William H. Rybolt, John D. McKenzie, Jr., Babson College, Robert N. Goldman, Simons College
William H. Rybolt, Babson College, Babson Park, MA 02157-0310

KEY WORDS: Excel, Graphs, Minitab, Regression

Abstract

In this paper a regression model for explaining variation in tuition at Massachusetts colleges and universities will be presented. The presentation will rely primarily on graphics obtained with Microsoft's Excel package and Minitab for Windows. The analysis is based on the data set from the 1995 Data Analysis Exposition, "What's What Among American Colleges and Universities?".

Problem

This report presents a primarily graphical approach to describing a parsimonious statistical model for explaining variation in tuition at Massachusetts colleges and universities. The data are a small part of the data set from the 1995 Data Analysis Exposition, "What's What Among American Colleges and Universities?"

We have had three aims in writing this report. The primary motivation was to demonstrate the extent to which modern graphics, if carefully used, can provide great insight into understanding a statistical model. A second aim was to compare the graphic capabilities of a dedicated statistics package such as Minitab Release 10 for Windows with those of spreadsheet package such as Excel 5.0. A final reason for this paper was to provide a helpful resource to the authors' respective colleagues. College administrators spend a good deal of time each year deciding on their schools' tuition. We hope the model derived in this paper will provide helpful insights for those administrators and their faculty and students.

Data

As mentioned above, the data for this project came from the 1995 Data Analysis Exposition's two data sets that were provided by U.S. News & World Report and the AAUP. Only the 56 four-year colleges in Massachusetts were included in this analysis.

Some adjustments were made to the original data. For example, a single measure for verbal SAT was obtained from the mean verbal SAT. If this value was not available, the verbal SAT midhinge (the mean of Q1 and Q3) was used. This latter substitution was justified by the closeness between the verbal SAT mean and midhinge values for colleges with both statistics.

There were only three schools for which neither mean verbal SAT scores or quartiles were available. The same process was repeated for math SAT scores. Combined SAT scores were the sum of the verbal and math SAT scores.

Another adjustment was to calculate two new statistics. pAccept was created by dividing the number of accepted students by the number of applicants. Similarly, pEnroll was computed by dividing the class size by the number of accepted students. Please refer to the Information on the Worksheet Exhibit for information about the 52 original variables and 13 additional variables.

The Tuition Variable

The aim of the analysis was to obtain a model relating out-of-state tuition to a small number of number of predictors selected from all of the variables in the augmented data set. (Out-of-state tuition was arbitrarily selected over in-state tuition, although subsequent analyses yielded similar results.) Please refer to the Data Display and Descriptive Statistics Exhibit for a listing of this tuition variable.

A histogram of the 56 tuition amounts is shown in the exhibit on page E3. From this display, it appears that the distribution of out-of-state tuition is approximately normal. Another informative display is presented in the Character Stem-and-Leaf Display exhibit on page E5. It reflects a more uniform distribution of tuition values. It also identifies 8 schools with relatively low tuitions.

Model Development Process

Due to too large a set of variables to place into a best subsets regression, high collinearity among many of the variables, and a large number of missing observations present in some variables, the following procedure was used to obtain our final model. First, possible predictors with too many missing observations were removed from consideration. Then a reasonable number of possible predictors were obtained by use of best subset regressions, multiple linear regressions, correlation coefficients, and subject matter knowledge. These predictors were placed into the Best Subsets Regression exhibit on page E7.

This output produced, with the assistance of the plot on page E8, models with two predictors. Due to the large number of missing observations present in the data, the actual model was then selected by the use of multiple linear regressions. Based upon the output from these regressions it was decided to use a model in which OTuition, out-of-state tuition, was the response variable and NPubPriv, an indicator variable indicating the school support, and VSAT, Verbal SAT, were the predictors variables. The codes in NPubPriv were 0 for public schools and 1 for private schools. (The choice of this model was confirmation by various stepwise regression and best subsets regression combinations.) Please refer to the output entitled Data Display for a complete listing of these variables.

The 3-D plot entitled Out-of-State Tuition vs. Verbal SAT and Public or Private is a graphical display of the these three variables. The exhibits on pages E12 and E13 present two-variable displays of the same data.

Final Model

A multiple linear regression analysis was produced; its output is present in the exhibit entitled Regression Analysis. This is an excellent model based on a perusal of the values of its s, R-sq, and VIF statistics and its four p-values. In addition, there are only two unusual observations. Both schools have unusually large standardized residuals. Administrators might consider one of the schools to be overpriced, while the other school is underpriced. None of the schools have unusually large leverage values. A graphical display illustrating this model is present in the exhibit entitled Best Model Relating Out-of-State Tuition to Verbal SAT and Public or Private.

Model Diagnostics

In the exhibit entitled Standardized Residual Model Diagnostics there are four common graphical displays to check standard linear regression assumptions. None of these displays indicated any problems with the assumptions. Each of these displays was individually recreated with at least one enhancement: Probability Plot of Standardized Residuals with its marginal boxplots, Histogram of Standardized Residuals with its cutpoint x-axis, I Chart of Standardized Residuals with its lack of problem observations based on eight tests, and Two Unusual Standardized Residuals with its explicit indication of public and private schools (and use of brushing to identify the school with the largest standardized residual). These displays are followed by six additional diagnostic plots: All Leverages are Less than .2, Cook's Distance vs. College, Largest Cook's Distance is Far Below F(3,50,.50). The t Distribution is a Good Fit to the Studentized Residuals, and The Ratio of MSE(i) and MSE vs. College. None of these indicated problems with the model. This is followed

by a display on page E27 that identifies outlying colleges. Finally, a matrix plot of the three variables was produced. Entitled Collinearity?, it raises doubts about the lack of collinearity in the data as indicated above in the Regression Analysis exhibit.

Prediction

In the exhibit on page E29 are relatively narrow 95% confidence intervals for the mean Out-of-State tuition and prediction intervals for an individual Out-of-State tuition value given specific values of the predictors. This high quality of prediction is further supported by the exhibit on page E32 which shows a strong linear association between the observed and model fitted values for the colleges from Connecticut. (The exhibits on pages E30 and E31 list the data used for this exhibit.)

A Comment on Excel

All of the above displays were produced by Minitab, Release 10 for Windows. Based on our work with the popular spreadsheet package, Excel 5.0, we have concluded that at the present time Excel 5.0 is not the preferred software for a graphical modeling exercise. Present in the exhibits are three Excel displays that illustrate how we came to this conclusion. The display entitled Excel Default Histogram shows Excel's default histogram (chart) setting which produces gaps between the histogram components. Also note its misleading location of its components' endpoints, and its inclusion of an interesting unnamed unidentified category on the extreme right. The Scatterplot may produce unexpected results because it uses the leftmost variable in the selection on the spreadsheet for the X axis even when you have indicated that a different variable should be used. The 3-D plot entitled Naive Excel 3-D Plot Type that was obtained by using Excel's default settings is really a display with three time series. The first is OTuition, the second is NPubPriv, and the third is VSAT. All of their values are all plotted on the same vertical axis. The naive attempt to label each axis is incorrect.

Summary

We had three goals in writing this report and constructing its related displays. The first reason was to demonstrate the extent to which modern graphics if, carefully used, can provide great insight into a statistical process such as model building. With the use of Minitab we have presented numerous examples to illustrate this aim. We have also presented examples to support our current belief that the graphical displays produced by a dedicated statistics package such as Minitab Release 10 for Windows is superior to those produced by a spreadsheet package such as Excel 5.0. Finally , we have shown how a model such as the one derived in this paper can provide helpful insights for administrators, faculty, and students.

Data Sources and Minitab Commands

In addition to the two provided data files, the following two magazines were used in the preparation of this paper:

March/April 1994 issue of <u>Academe, Bulletin of the American Association of University Professors</u> <u>1995 College Guide of U.S. News & World Report's Best Colleges</u>

Copies of the commands that were used to generate the Minitab exhibits are available upon request from the contact author. He will provide information on the construction of the Excel exhibits. A complete set of exhibits is also available from him. (The exhibits denoted by a * are included on the next three pages.)

Exhibits

E3

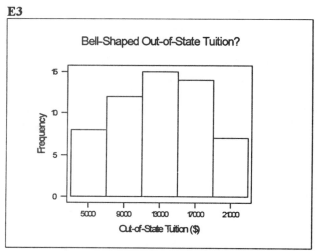

Bell-Shaped Out-of-State Tuition?

E4

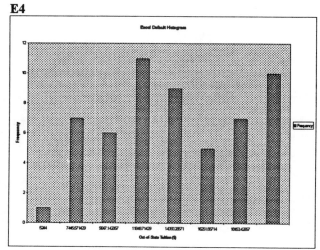

E8

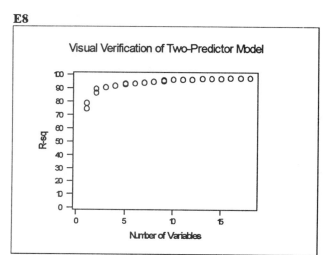

Visual Verification of Two-Predictor Model

E10

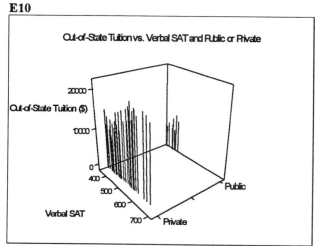

Out-of-State Tuition vs. Verbal SAT and Public or Private

E11

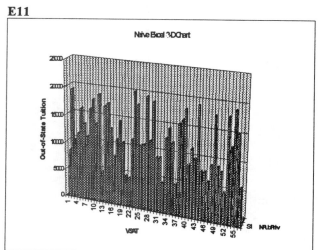

E13

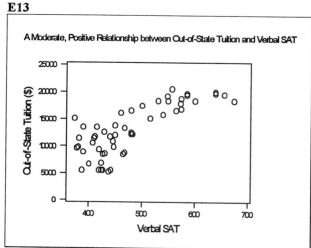

E14

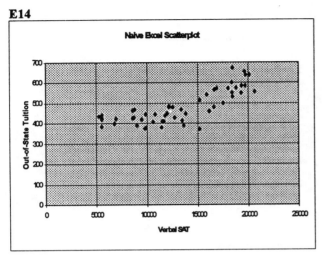

E16

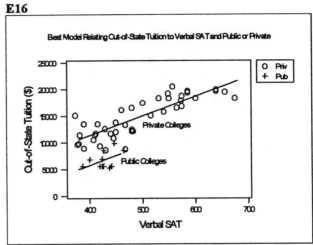

E18

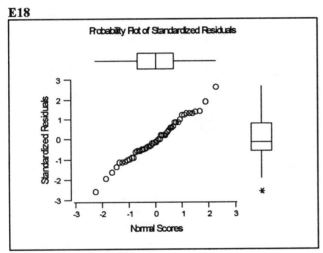

E19

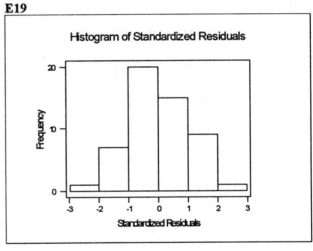

E20

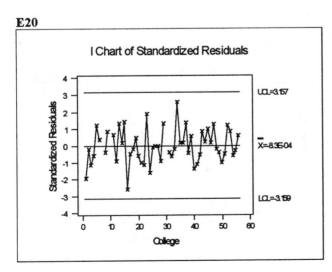

E21

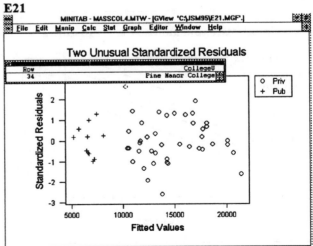

87

E24

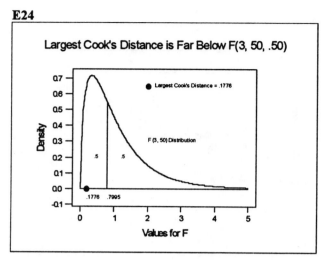

Largest Cook's Distance is Far Below F(3, 50, .50)

E25

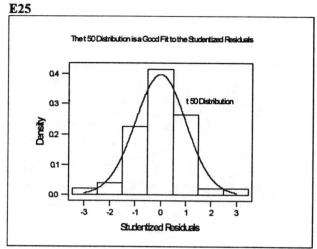

The t 50 Distribution is a Good Fit to the Studentized Residuals

E26

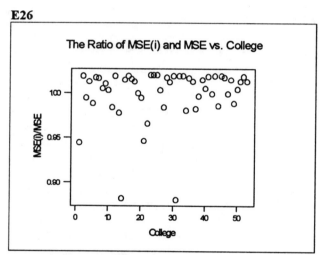

The Ratio of MSE(i) and MSE vs. College

E27

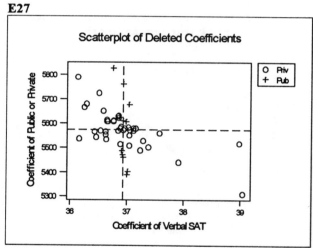

Scatterplot of Deleted Coefficients

E29

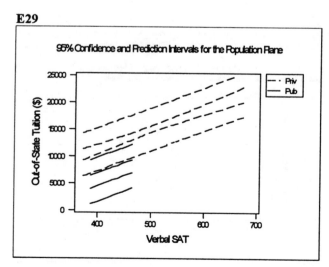

95% Confidence and Prediction Intervals for the Population Plane

E32

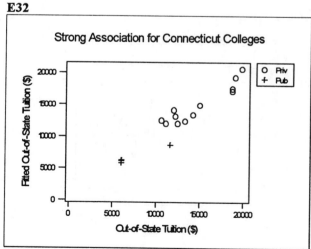

Strong Association for Connecticut Colleges

USING INFORMATION ON COLLEGESTO MODEL TUITION

William C. Rinaman, Lifang Hsu, J. Adam O'Hara
Le Moyne College, Syracuse, NY 13214

Key Words: Regression, Discriminant Analysis, Factor Analysis, Correlation

ABSTRACT

The US News and World Report data regarding colleges and universities in the United States are analyzed to try to determine what factors contribute to the amount of tuition charged by a school. Stepwise regression analysis is used to determine which variables from the survey best account for the variability observed in tuition costs. A linear discriminant function is derived to ascertain how well college data can determine if a school can be placed in the appropriate third for tuition costs. A factor analysis is done to try to determine factors that may be related to school quality. The major contributors to these factors are used to determine the degree to which they are correlated with tuition costs.

I. INTRODUCTION

Many different organizations publish ratings and other information on colleges and universities in the United States. These provide much useful information for prospective Freshmen and their families. Such data, while being useful, do not give a complete picture of an academic institution. In an effort to determine what can be derived from such data a statistical analysis was done to see how well information about a college would permit the prediction of tuition costs by consumers. Such a prediction would alert consumers to schools that may appear to be charging tuition significantly different from a typical institution having the same characteristics. In addition, an attempt was made to determine factors that relate to the "quality" of a school. An analysis was also done to determine if there is a correlation between a measure of "quality" and tuition.

II. METHODOLOGY

The population for the study consisted of 1302 colleges and universities in the United States. The data were collected by US News and World Report. The data set contained observations of the following attributes of each school--Federal ID number, college name, state, public/private, averages math SAT score, average verbal SAT score, average combined SAT score, first quartile math SAT score, third quartile math SAT score, first quartile verbal SAT score, third quartile verbal SAT score, average ACT score, first quartile ACT score, third quartile ACT score, number of applications received, number of applicants accepted, number of new students enrolled, percent new students from the top 10% of their high school class, percent new students from the top 25% of their high school class, number of full time undergraduates, number of part time undergraduates, in-state tuition, out-of-state tuition, room and board costs, room costs, board costs, additional fees, estimated book costs, estimated personal spending, percent of faculty with Ph.D.'s, percent of faculty with terminal degrees, student/faculty ratio, percent of alumni who donate, instructional expenditure per student, and graduation rate. Two additional variables were computed. These were the proportion of applicants that were accepted and the proportion of accepted applicants that enrolled. It was felt that these ratios were more meaningful than would be the raw data regarding number of applications received, number of applicants accepted and number of accepted applicants that enrolled.

Not very many schools reported on all of the test scores. Some schools reported only SAT scores while others provided only ACT scores. Within those groups some schools reported total scores while others gave only quartile data. This variability in test score reporting generated a large number of missing values even though schools did report test results. It was decided to reduce the number of missing values in test score reporting by using quartile data to predict total scores in the following manner. For those schools reporting only quartile values for each of the SAT tests the average combined SAT score was estimated by the following regression equation.

$$AVCSAT = 60.1 + 0.4705(MQ1 + MQ3 + VQ1 + VQ3),$$

where $AVCSAT$ = predicted average combined SAT score, $MQ1$ = first quartile math SAT score, $MQ3$ = third quartile math SAT score, $VQ1$ = first quartile verbal SAT score, and $VQ3$ = third quartile SAT score. A similar process was used to reduce the number of missing values for ACT scores. The regression equation used in that case was as follows.

$$AVACT = 3 + 0.436(ACTQ1 + ACTQ3),$$

where $AVACT$ = predicted average ACT score, $ACTQ1$ = first quartile ACT score, and $ACTQ3$ = third quartile ACT score.

It was decided that only those variables not related to expenditures by students would be used in the attempt to model tuition. This is because the purpose of this analysis was to see how well characteristics of the student body, faculty and administration determine the amount of tuition. In addition, the out-of-state tuition charged by public institutions was used. It was felt that this figure would best eliminate much variability due to the amount of state support for a school.

Scatter diagrams for all of the potential independent variables against out-of-state tuition were plotted to determine whether the assumption of a linear relationship was justified. After this was verified stepwise regressions were conducted. One used SAT scores while the other used ACT scores. Residual plots and normal plots were used to verify the regression model assumptions. In addition to the regression analysis, a linear discriminant analysis was conducted to determine the ability to use the independent variables to predict which third for out-of-state tuition each school should be assigned.

A second area of interest was that of measuring the "quality" of a school. Factor analyses were run to ascertain if the variability in the data could be attributed to factors that might reflective differences in school "quality." The attributes of student level, quality of faculty and what might be termed "image" were found. These were correlated against out-of-state tuition to see if there appeared to be any relationship between these quality factors and the amount of tuition charged by a given school.

All analyses were done in Minitab Release 10 for Windows. Graphics were also produced in Minitab. The variables used in the analyses are defined as follows.

SECTOR = public (1) or private (2)
OUTTUIT = out-of-state tuition
PCT10 = percent from top 10 % of h.s. class
PCT25 = percent from top 25% of h.s. class
PCTPHD = percent of faculty with Ph.D.'s

PCTTERM = percent of faculty with terminal deg.
FTUG = number of full-time undergraduates
SFRATIO = student/faculty ratio
PCTDON = percent of alumni who donate
INS_EXP = instructional expenditure/student
G_RATE = graduation rate
SAT = combined SAT score
ACT = ACT score
ACRC = prop. of applicants that are accepted
ENRAC = prop. of accepted applicants who enroll

IV. RESULTS
A. Modeling Tuition Costs

Two stepwise regressions using forward selection with an F-to-enter value of 4.0 were conducted to obtain a set of independent variables to predict out-of-state tuition. They differed only in that one used SAT scores and the other used ACT scores. The final regression using SAT scores showed the following results. The regression equation is

OUTTUIT = - 8334 + 0.172 INS_EXP
+ 3379 SECTOR + 11.1 SAT
- 5183 ENRAC + 31.9 PCTDON
+ 33.4 PCTTERM - 27.6 PCT10
- 74.8 SFRATIO + 16.0 G_RATE.

Out of the 1302 schools in the data base 702 were used and 600 contained missing values. The analysis showed that all coefficients were significant at the .01 level or better. The overall F statistic was highly significant. In addition, the R^2 value was 0.742. The residual plot showed no evidence of an unaccounted for non-linear trend. The residual plot and the normal probability plot are shown in Figure 1 and Figure 2, respectively.

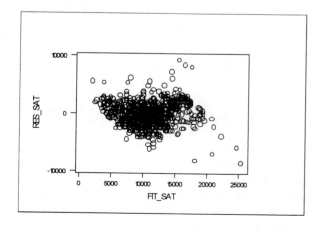

Figure 1.

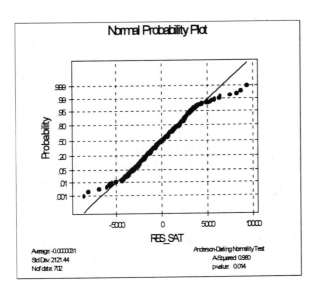

Figure 2.

OUTTUIT = - 4902 + 0.265 INS_EXP
    + 2697 SECTOR + 202.4 ACT
    - 4728.4 ENRAC
    + 41.2 PCTTERM
    + 25.9 G_RATE + 24.7 PCTDON
    - 0.043 FTUG.

In this case 681 schools contained no missing values. The summary statistics for the individual coefficients and the overall model were similar to those for the previous analysis. This time the R^2 value was 0.743. The residual and normal plots are shown in Figure 3 and Figure 4, respectively.

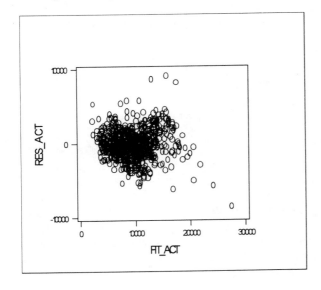

Figure 3.

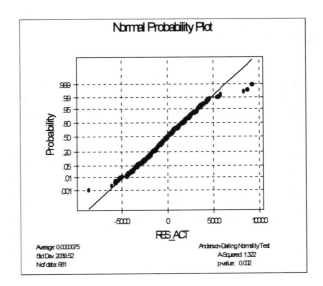

Figure 4.

As was the case for the analysis using SAT scores, the residuals appear to be fairly random and somewhat leptokurtic.

These analyses indicate that it is possible to account for a fair degree of the variability in tuition costs by considering the non-monetary variables provided by the US New and World Report survey. There do appear to be, however, a number of other factors that go into the setting of school tuition that are not reflected in these data.

It was also of interest to determine if linear discriminant functions could successfully use non-monetary information to classify according to tuition levels. For this purpose the 1302 schools were divided into four groups according to whether their out-of-state tuition placed each either in the bottom third (Group 1), the middle third(Group 2), or in the top third(Group 3). A linear discriminant analysis yielded the following results. The results are first shown for schools reporting SAT scores.

	Actual Group		
Put Into Group	1	2	3
1	94	55	7
2	24	128	54
3	4	36	287
Total	122	219	348
No. Correct	94	128	287
Proportion	0.770	0.584	0.825

N = 689   N Correct = 509   Prop. Correct = 0.739

The results for schools reporting ACT scores are as follows.

	Actual Group		
Put Into Group	1	2	3
1	106	55	2
2	35	132	40
3	0	32	193
Total	141	219	235
No. Correct	106	132	193
Proportion	0.752	0.603	0.821

N = 595   N Correct = 431   Prop. Correct = 0.724

In both cases the ability to correctly classify schools belonging Group II was the poorest.

The discriminant analyses support the conclusions reached using the regression results. That is, the data provided permits a reasonable degree of ability to determine which tuition category a school should be assigned to based on non-monetary data. However, a number of additional factors appear to be working when schools determine tuition rates.

In comparing the two types of analysis the following can be observed. Not surprisingly, both methods assigned high importance to whether a school was public or private, student test scores, percentage of faculty with terminal degrees and proportion of students that are accepted that actually enroll. On the other hand instructional expenses per student was the first variable to be included by the stepwise regression procedures, but it was given essentially no weight by the discriminant functions. The percentage of alumni who donate was given positive weight in the regression equations and a negative weight in the discriminant functions. For the most part the signs of the coefficients make sense on an intuitive basis. One exception is the negative sign both regressions gave to the proportion of accepted students who actually enroll. Another is the positive sign the discriminant analyses assigned to the student/faculty ratio.

B. Measuring School Quality

To ascertain the degree to which the variability in the non-monetary information could be attributed to the "quality" of an institution a factor analysis using a principal factor extraction with a varimax rotation for four factors was conducted. The loadings suggest the following interpretations. Factor 1, exhibited high positive loadings to PCT10, PCT25, SAT and ACT along with the high negative loading to ACRC. Thus, it appears to reflect the selectivity of the school with regard to prospective undergraduates or the "quality" of the student body. Factor 2 appears to reflect some attribute about the "quality" of the faculty because it gave high negative loadings on PCTPHD and PCTTERM. Factor 3 gave high negative loadings on FTUG and SFRATIO. This supports an interpretation that Factor 3 reflects the size of the institution. It is questionable whether that reflects on the "quality" of a school. Factor 4, had a high positive loading on ENRAC and high negative loading on G_RATE. It has been interpreted to reflect the desire of students to attend a school and the support a school gives to help students succeed. This can be taken to reflect the image of "quality" that is perceived by prospective students. These factors led to the consideration of the following three measures of institutional quality. The constant of 1 and the divisors are used to have the measures to range between 0 and 1.

Student quality:
1) SAT scores
$$SQS = [1 + (PCT10 + PCT25)/200 + SAT/1600 - ACRC]/3$$
2) ACT scores
$$SQA = [1 + (PCT10 + PCT25)/200 + ACT/36 - ACRC]/3$$

Faculty quality:
$$FQ = (PCTPHD + PCTTERM)/200$$

Image:
$$IM = [1 + G\_RATE/100 - ENRAC]/2$$

Total:
1) SAT scores
$$TQS = [SQS + FQ + IM]/3$$
2) ACT scores
$$TQA = [SQA + FQ + IM]/3$$

These measures were correlated with out-of-state tuition to see the degree to which each measure was related to the cost of attendance. SQS had a correlation of 0.554 with tuition, SQA had a correlation of 0.539 with tuition, FQ had a correlation of 0.407 with tuition, IM had a correlation of 0.655 with tuition, TQS had a correlation of 0.657 with tuition and TQA had a correlation of 0.676 with tuition.

It was also of interest to see how the measure of total quality was related to out-of-state tuition costs. Thus, the out-of-state tuition costs were divided by the maximum value to create a variable whose range was between 0 and 1. This was then divided into the total quality measures. These measures, along with the other quality measures were computed for each of the schools for which there were no missing values. The schools were then ranked according to these measures. The rankings were different than when tuition costs were not taken into account.

CHOOSE THE RIGHT SCHOOL: SOME INVESTIGATION ABOUT THE QUALITY OF EDUCATION, TUITION AND FACULTY SALARY STRUCTURE

Nan-Jung Hsu, Hsin-Cheng Huang, Pradipta Sarkar, Kevin Wright, Iowa State University
Pradipta Sarkar, 308 Snedecor Hall, Iowa State University, Ames, Iowa 50011

Key Words : Cluster Analysis, Data Imputation, Exploratory Data Analysis.

1 Introduction

In this short paper, we are interested in three basic questions related to the US undergraduate education. First, the quality of students and teachers in the U.S. colleges/universities; second, if the tuition is worth paying — are the students getting what they are paying for; and third, the faculty salary structure. In the following sections we briefly discuss the methods followed and then summarize the main findings. We have come up with a ranking of all the U.S. schools using different criteria. Details of the findings has been demonstrated in the poster session with the help of some graphs, pictures and diagrams. We use XGobi for high dimensional exploratory data analysis and SAS and S-Plus for statistical computations.

2 Findings

2.1 Quality of Students and Teachers

To measure the quality of students we use the average scores of the entering undergraduates in SAT and ACT. If both SAT and ACT scores are available for a school, the ACT score is converted to an equivalent SAT score by regressing the average SAT score on the average ACT score. The real SAT and "pseudo-SAT" scores are then averaged to produce an index of standardized test scores. If only the data on the ACT scores is available, the equivalent SAT score is imputed using the estimated value from the regression of the average SAT score on the average ACT score and this equivalent SAT score is used as an index of the standardized test score. On the basis of this index the top 5 schools are California Institute of Technology (CA), Harvard University (MA),

Harvey Mudd College (CA), Princeton University (NJ) and Stanford University (CA).

Another index of the quality of entering students is the percentage of students from the top 10% (or top 25%) of the H.S. class. So the schools are ranked according to the percentage of entering students from the top 10% and the top 25% of the H.S. class also. California Institute of Technology (CA), MIT (MA), different campuses of University of California (San Diego, Santa Cruz, Los Angeles, Riverside, Davis, Berkeley, Santa Barbara, Irvine), Yale University (CT), Harvey Mudd College (CA), Harvard University (MA), Cooper Union (NY), SUNY Buffalo (NY) rank very high according to these criteria.

Percentage of faculty with PhD degree can serve as an index of the quality of the teachers although it might not as important as it is for the graduate school. With the help of a dot plot in Xgobi the following schools were identified as having almost 100% PhD faculty : Bryn Mawr College (PA), Kings College (NY), California Institute of Technology (CA), Pitzar College (CA), University of Judaism (CA), CUNY - Queens College (NY), Harvey Mudd College (CA), University of California (CA).

2.2 Quality of Education

The concept of the quality of education is rather subjective and very difficult to measure quantitatively. In an attempt to quantify it, the following variables are chosen : student to faculty ratio, instructional expenditure per student, pct. of alumni who donate, pct. of faculty with PhD and graduation rate. We compute an index of the quality of education by first standardizing the variables and then assigning different weights to these variables and taking the corresponding linear combination. Sensitivity analysis on the standardized variables is done by changing the weights. Certain schools show up in the top 10 again and again even when changing the relative weights of the variables within a wide range. We conclude that these are certainly the best schools as they are robust to changes in the relative weights of the variables of interest. California Institute of Technology (CA), Johns Hopkins University (MD), Yale University (CT), Wake Forest University (NC), Washington University (MO), University of Chicago (IL), Princeton University (NJ) are among the schools which turn up in the top 10 most of the time.

One important question associated with the quality of education is tuition. Is the tuition worth paying? Tuition is dependent on the type of the school (private/public). Generally the public schools have low in-state tuition. On the other hand, for the private schools the in-state and the out-of-state tuition are generally equal. Also in a public school most of the students are from that state itself. So for the public schools the in-state tuition serves as a good measure of "how much the student is paying." We identified the schools with low tuition but higher quality of education. These are the schools that one would prefer so far as the "return out of tuition" is concerned. Due to page limitation we cannot present them here. A list of such schools is available from the authors.

2.3 Faculty Salary

Private schools have more variation in the faculty salary and compensation structure than the public ones. A scatter plot of the number of full-time students vs. average salary (alternatively average compensation), public and private schools being brushed with two different colors, explains this fact clearly. This plot indicates that the public schools pay moderately and have high student enrolment. On either extreme of the average salary (average compensation) scale are mostly the private schools. The ones on the higher side include, as expected, famous schools like California Institute of Technology (CA), Stanford University (CA), Harvard University (MA), MIT (MA), Princeton University (NJ) etc. Interestingly these are the schools with high teacher to student ratio too. On the lower side of the salary (compensation) scale there are some small private colleges or universities. The ones in the lowest range of faculty salary include Tabor College (KS), Colorado Christian University (Co), Dakota Wesleyan University (SD), Benedict College (SC), St. Mary-of-the-Woods College (IN) and Voorhees College (SC). Faculty salary also depends on the region. So we do not find many low salary schools in the expensive places.

Iowa State University ranks in the top 1/3, which is an indication that it pays well considering that Iowa has a low cost of living.

2.4 Cluster Analysis

Overall rank of a school is determined by the quality of students, quality of teachers and quality of education. We calculate an index of overall rank using the variables used for quality of students, quality of teachers and quality of education. Theoretically speaking each variable can take values from $-\infty$ to $+\infty$, and if the value of a certain variable is either too big or too small, the indices that use that variable will be very much sensitive to that variable. In order to keep a check on that we truncate the distribution of the variables both sides so that the standardized variables take values in the range -5 to 5. A list of the overall ranks is available from the authors. All the schools are grouped into 20 different clusters using Ward's method. With the help of Xgobi, we try to find out the different characteristics associated with the clusters and discovered some intersting results. Some of the findings are summarized below.

- Clusters with top schools are smaller in size.

- The cluster consisting of 7 schools with overall rankings 1,2,4,5,6,8 and 9 (See Table 1 for names) looks like the best cluster. For these schools the quality of education index is better than the quality of students index.

- Cluster #1 has 21 schools with the overall rank in the range 12-50. It has 4 schools in the top 20. Except for University of Pennsylvania the other three top 20 schools have better index for the quality of students than that of education.

- We have got a few very big clusters consisting of low rank but inexpensive schools.

- Military schools have high expenditure per student which is responsible for very high quality of education indices and in turn high (to a lesser extent) overall rank. Their quality of students indices are not so high.

- There are two clusters (viz, #18 & #19) each with 11 schools of comparable overall rankings; for the first group the quality of students is far better than the quality of education and for the second one exacly the opposite.

To get a better visual comparison of the cluster characteristics one can select certain variables of interest and then randomly select a few schools from

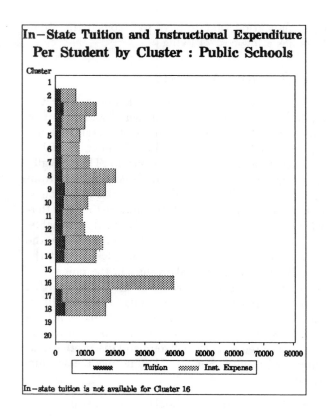

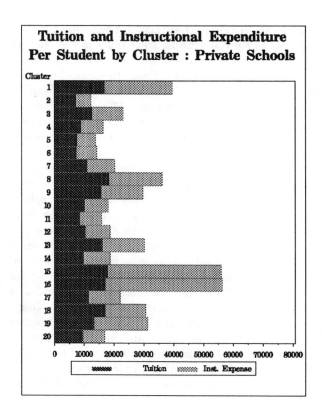

Figure 1: Clusterwise Comparison of Tuition

Table 1: Top 20 Schools (Overall Ranking for Undergraduate Education)

Name of the School	State	Name of the School	State
1. Johns Hopkins University	MD	11. Harvard University	MA
2. Wake Forest University	NC	12. University of Pennsylvania	PA
3. Yale University	CT	13. Massachusetts Institute of Technology	MA
4. Washington University	MO	14. Case Western Reserve University	OH
5. University of Chicago	IL	15. Princeton University	NJ
6. Dartmouth College	NH	16. Stanford University	CA
7. California Institute of Technology	CA	17. United States Air Force Academy	CO
8. Duke University	NC	18. Northwestern University	IL
9. Emory University	GA	19. United States Naval Academy	MD
10. Columbia University	NY	20. Harvey Mudd College	CA

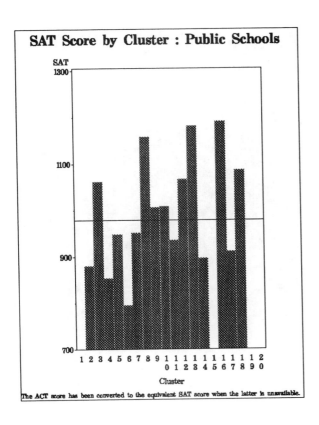

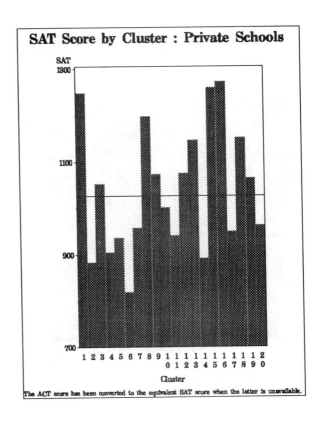

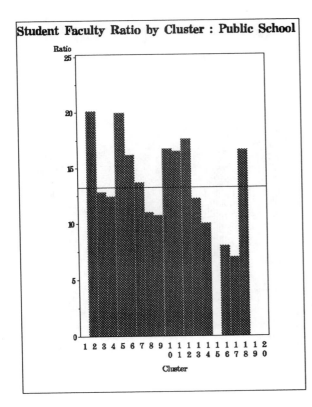

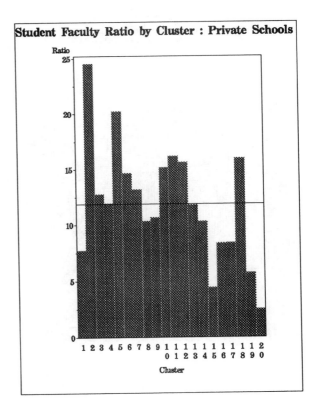

Figure 2: Clusterwise Comparison of Average SAT Scores and Student-Faculty Ratio

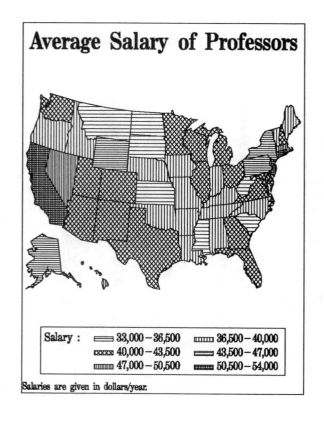

Average Salary of Professors

Salary : ▭ 33,000–36,500   ▥ 36,500–40,000
         ▧ 40,000–43,500   ▤ 43,500–47,000
         ▨ 47,000–50,500   ▰ 50,500–54,000

Salaries are given in dollars/year.

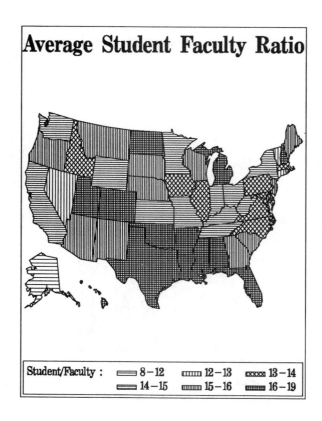

Average Student Faculty Ratio

Student/Faculty : ▭ 8–12   ▥ 12–13   ▧ 13–14
                  ▤ 14–15   ▨ 15–16   ▰ 16–19

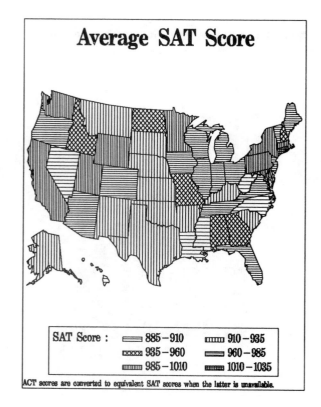

Average SAT Score

SAT Score : ▭ 885–910   ▥ 910–935
            ▧ 935–960   ▤ 960–985
            ▨ 985–1010 ▰ 1010–1035

ACT scores are converted to equivalent SAT scores when the latter is unavailable.

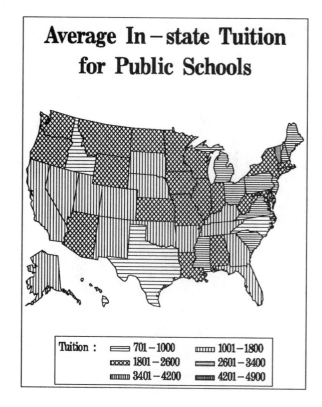

Average In–state Tuition for Public Schools

Tuition : ▭ 701–1000   ▥ 1001–1800
          ▧ 1801–2600 ▤ 2601–3400
          ▨ 3401–4200 ▰ 4201–4900

Figure 3: Regional Analysis of Certain Variables of Interest

each cluster, plot these variables for the selected schools and finally join the schools of different clusters by straight lines of different colors. Because of limitation of space we are not displaying such graphs here. A complete list of the clusters is also available from the authors.

Figure 1 shows that the average instate tuition for the public schools does not vary much across the clusters. Figure 2 gives a clusterwise comparison of the average SAT scores and the student to faculty ratio, separately for the private and the public schools. The median values are indicated by a solid line in each of these pictures.

2.5 Statewise Comparison

Statewise averages of certain variables, like faculty salary, average SAT scores, student-faculty ratio and instate tuition for the public schools, plotted on the map of USA is very useful to have a snapshot of the schools all over the nation (See figure 3).

2.6 Iowa Schools

As per our analysis Grinnel College is the best undergraduate school in the state of Iowa with an overall ranking of 49. Interestingly both the quality of education and the quality of students indices for Grinnel College are 60. Iowa State University, Luther College, Cornell College, and University of Iowa are among other good schools in the state.

3 Conclusion

Choosing the right school is a difficult problem to solve and it depends on what one's priorities are. Often, rankings of the schools are available in different magazines and journals. One should not get confused by such rankings, since different organizations use different criteria for assigning ranks. Rankings are just one of the several criteria a prospective student should take into account in choosing a school. Simply because a school is at the top of its category does not automatically mean that it is the best choice for a particular student. Apart from rankings there are many other factors, such as the students' academic and professional goal, financial constraints, location of the school, facilities and the size of the school etc., that play important roles in the choice of an appropriate school. Also it is a well established fact that the name of a reputed school on a resume plays a big role in getting a good job. So ideally the reputation of the school should be a factor in determining the rank. An index of reputation can be calculated by asking questions to a true representative sample of the population. But this factor is absent in our analysis. We refrained ourselves from using the reputation of the school as a factor to calculate the rank because it is a very much subjective matter and it is very difficult for us to get a sample of people that is a true representative of the US population.

Our suggested rating (complete list is available from the authors on request) may be used as a guideline to choose the school, but always keeping the above mentioned points in mind and also remembering that a small difference in the ranking does not necessarily make two schools qualitatively much different from each other.

Acknowledgements

We would like to thank Pankaj Srivastava of the College of Business, Shiping Liu of the Dept. of Agricultural Economics and Huaichin Chen of the Dept. of Statistics, Iowa State University for helping us in several ways to make this project successful.

Index of Participants